Chance and Necessity

JACQUES MONOD, together with André Lwoff and
François Jacob, was awarded the Nobel Prize in 1965
for elucidating the replication mechanism of genetic
material and the manner in which cells synthesize
protein.

Dr Monod is now the director of the Pasteur Institute
in Paris, whose Cellular Biochemistry Service he
created in 1954 and directed thereafter. He was
appointed a professor at the College of France in
1967, and he is a foreign member of both the Royal
Society and the National Academy of Sciences.

Jacques Monod

Chance and Necessity

An Essay on the Natural Philosophy
of Modern Biology

Translated from the French by Austryn Wainhouse

COLLINS/FOUNT PAPERBACKS

This book was originally published in France in 1970 under
the title *Le Hasard et la nécessité* by Éditions du Seuil, Paris

© Éditions du Seuil 1970
English translation copyright © 1971 by Alfred A. Knopf, Inc.

First published in Great Britain
by William Collins Sons & Co Ltd 1972
First issued in Fontana Books 1974
Reprinted in Fount Paperbacks March 1977
Fourth Impression October 1983

Made and printed in Great Britain
William Collins Sons & Co Ltd Glasgow

*Everything existing in the Universe is the fruit
of chance and of necessity.*

DEMOCRITUS

At that subtle moment when man glances backward over his life,
Sisyphus returning towards his rock, in that slight pivoting, he
contemplates that series of unrelated actions which becomes his
fate, created by him, combined under his memory's eye and soon
sealed by his death. Thus, convinced of the wholly human origin
of all that is human, a blind man eager to see, who knows that the
night has no end, he is still on the go. The rock is still rolling.

I leave Sisyphus at the foot of the mountain! One always finds
one's burden again. But Sisyphus teaches the higher fidelity that
negates the gods and raises rocks. He, too, concludes that all is
well. This universe henceforth without a master seems to him
neither sterile nor futile. Each atom of that stone, each mineral
flake of that night-filled mountain, in itself forms a world. The
struggle itself towards the heights is enough to fill a man's heart.
One must imagine Sisyphus happy.

ALBERT CAMUS, *The Myth of Sisyphus*

Contents

the central nervous system – the analysis of sense impressions –
empiricism and innateness – the function of simulation – the
dualist illusion and the presence of the spirit

Preface

Biology occupies a position among the sciences both marginal and central. Marginal because, the living world constituting only a tiny and very 'special' part of the universe, it does not seem likely that the study of living beings will ever uncover general laws applicable outside the biosphere. But if the ultimate aim of the whole of science is indeed, as I believe, to clarify man's relationship to the universe, then biology must be accorded a central position, since of all the disciplines it is the one that endeavours to go most directly to the heart of the problems that must be resolved before that of 'human nature' can even be framed in other than metaphysical terms.

Consequently no other science has quite the same significance for man; none has already contributed so greatly to the shaping of modern thought, profoundly and definitively affected as it has been in every domain, philosophical, religious, political, by the advent of the theory of evolution. Its phenomenological validity generally accepted by the close of the last century, the theory of evolution, while dominating the whole of biology, still remained as if suspended, awaiting the elaboration of a *physical* theory of heredity. Thirty years ago the hope that one would soon be forthcoming appeared almost illusory, notwithstanding the successes in classical genetics. Today, however, this is precisely what we have in the molecular theory of the genetic code. Here 'theory of the genetic code' is to be understood in the broader sense, including not only concepts relevant to the chemical structure of hereditary material and the information it conveys, but also the molecular mechanisms for expressing this information morphogenetically and physiologically. So defined, the theory of the genetic code constitutes the fundamental basis of biology. This does not mean, of course, that the complex structures and functions of organisms can be *deduced* from it, nor even that they are always directly analysable on the molecular level. (Nor can everything in chemistry be predicted or resolved by means of the quantum theory which, without question, underlies all chemistry.)

But although the molecular theory of the genetic code cannot now – and will doubtless never be able to – predict and resolve the whole of the biosphere, it does today constitute a general theory of living systems. No such thing existed in scientific knowledge before molecular biology. Until then the 'secret of life' seemed to be essentially inaccessible. In recent times much of it has been laid bare. This, a considerable event, should certainly deeply affect contemporary thinking, once the general significance and consequences of the theory are understood and appreciated beyond the narrow circle of specialists. I hope this essay may contribute to that end. In it, rather than making a thorough survey of the contents of modern thinking in biology, I have tried to bring out the 'form' of its key concepts and to point out their logical relationships with other areas of thought.

Nowadays it is risky for a man of science to use the word *philosophy*, even with the qualification *natural*, in the title (or even subtitle) of a book: this is guaranteed to earn it a distrustful reception from other scientists, and from philosophers at best a condescending one. I have only one excuse, but I believe it is sound: the duty, today more imperative than ever, which is incumbent on scientists to consider their discipline within the larger framework of modern culture, with a view to enriching the latter not only with technically important findings, but also with what they may feel to be humanly significant ideas arising from their area of special concern. The very ingenuousness of a fresh look at things (and science possesses an ever-youthful eye) may sometimes shed a new light upon old problems.

Meanwhile, to be sure, any confusion between the ideas *suggested* by science and science itself must be carefully avoided; but it is just as necessary that scientifically warranted conclusions be resolutely pursued to the point where their full meaning becomes clear. A difficult exercise. I do not claim to have given a faultless performance. First let me say that in what follows the strictly biological part is in no sense original: I have done no more than summarize what are considered established ideas in contemporary science. It is true that the relative importance given to different developments, like the choice of examples offered, reflects personal tendencies. Some outstanding chapters in biology are not even mentioned. But this essay does not seek to discuss the entire field of biology; it is an attempt to extract the quintessence of the molecular theory of the code. The ideological generalizations I have ventured to deduce from it are of course my sole responsibility. But I believe that, where they do not exceed the bounds of epistemology, these interpretations would be accepted by the majority of modern biologists. I must claim full responsibility also for the ethical and sometimes political ideas expressed, which I have preferred not to avoid, perilous though they are and however naïve or overambitious they may appear. The scientist should be modest but not at the expense of his convictions, which he *must* defend. There again, however, I have the encouraging certainty of finding myself in full agreement with certain contemporary biologists whose achievements are worthy of the highest respect.

I must beg the indulgence of biologists for passages which will strike them as tediously self-evident explanation; of nonbiologists for the dryness of other pages given over to unavoidable 'technical' background. Some readers may be helped over these difficulties by the appendices. But I should like to stress that they can be dispensed with by anyone who is not disposed to grapple directly with the chemical realities of biology.

This essay is the result of a series of lectures – the Robbins Lectures – given in February of 1969 at Pomona College in California. I wish to thank the authorities of that college for having provided me with the occasion to explore, before a very young and eager audience, certain themes which had for a long time been with me a subject for thought, but not for teaching. These themes were also at the core of a course I gave at the Collège de France during the 1969–70 academic year. It is a fine and precious institution that allows its members sometimes to step beyond the strict boundaries of their charge and purview. Thanks therefore be unto Guillaume Budé and King Francis I.

April 1970   CLOS SAINT-JACQUES

1

About Strange Objects

The difference between artificial and natural objects appears immediate and unambiguous to all of us. A rock, a mountain, a river, or a cloud – these are natural objects; a knife, a handkerchief, a car – are artificial objects, artifacts.[1] Analyse these judgments, however, and it will be seen that they are neither immediate nor strictly objective. We know that the knife was man-made for a use its maker visualized beforehand. The object renders in material form the pre-existent intention that gave birth to it, and its form is explained by the performance expected of it even before it takes shape. It is another story altogether with the river or the rock which we know, or believe, to have been moulded by the free play of physical forces to which we cannot attribute any design, any 'project' or purpose. Not, that is, if we accept the basic premise of the scientific method, to wit, that nature is *objective* and not *projective*.

Hence it is through reference to our own activity, conscious and projective, intentional and purposive – it is as makers of artifacts – that we judge of a given object's 'naturalness' or 'artificialness'. Might there be objective and general standards for defining the characteristics of artificial objects, products of a conscious purposive activity, as against natural objects,

1. In the literal sense: products of human art or workmanship.

resulting from the gratuitous play of physical forces? To be sure of the complete objectivity of the criteria chosen, it would doubtless be best to ask oneself whether, using these criteria, a programme could be drawn up enabling a computer to distinguish an artifact from a natural object.

Such a programme could have the most interesting applications. Let us suppose that a spacecraft is soon to be landed upon Venus or Mars; what could be more fascinating than to find out whether our neighbouring planets are, or at some earlier period ever were, inhabited by intelligent beings capable of projective activity? In order to detect such present or past activity we would have to search for and be able to recognize its *products*, however radically unlike the fruit of human industry they might be. Wholly ignorant of the nature of such beings and of the projects they might have conceived, our programme would have to use only very general criteria, solely based on the structure and form of the objects examined and without any bearing upon their eventual function.

The suitable criteria, we see, would be two in number: (a) regularity, and (b) repetition. By means of the first one would seek to make use of the fact that natural objects, wrought by the play of physical forces, almost never present geometrically simple structures such as flat surfaces, rectilinear edges, right angles, or exact symmetries; whereas artifacts will ordinarily show such features, if only in an approximate or rudimentary manner.

Of the two criteria, repetition would probably be the more decisive. Materializing a repeated plan, homologous artifacts meant for the same use reflect more or less faithfully the constant purpose of their creator. In that respect the discovery of numerous specimens of closely similar objects would be very significant.

These, briefly defined, are the general criteria that might serve. It must be added that the objects selected for examination would be of *macroscopic* dimensions, but not *microscopic*.

By macroscopic is meant dimensions measurable, say, in centimetres; by microscopic, dimensions normally expressed in angstroms (a hundred million of which equal one centimetre). This proviso is crucial, for on the microscopic scale one would be dealing with atomic and molecular structures whose simple and repetitive geometries, obviously, would attest not to a conscious and rational intention but to the laws of chemistry.

Now let us suppose the programme drawn up and the machine built. To check its performance, the best possible test would be to try it out on terrestrial objects. Let us invert *difficulties of a space programme* our hypotheses and imagine that the machine has been put together by the experts of a Martian NASA attempting to detect evidence of organized, artifact-producing activity on Earth. And let us suppose that the first Martian craft comes down in the Forest of Fontainebleau, not far, let's say, from the village of Barbizon. The machine views and compares the two series of objects most prominent in the area: on the one hand the houses in Barbizon, on the other hand the rock formations of Apremont. Using the criteria of regularity, of geometric simplicity, and of repetition, it will easily decide that the rocks are natural objects and the houses artifacts.

Proceeding to focus on lesser objects, the machine examines some pebbles, near which it discovers some crystals – quartz crystals possibly. According to the same criteria it should of course decide that while the pebbles are natural, the quartz crystals are artificial objects. This decision would appear to point to some 'error' in the writing of the programme, an 'error' which, moreover, proceeds from an interesting source: if the crystals present perfectly defined geometrical shapes, that is because their macroscopic structure directly reflects the simple and repetitive microscopic structure of the atoms or molecules constituting them. A crystal, in other words, is the macroscopic expression of a microscopic structure. This 'error', incidentally, should be easy enough to eliminate, since all *possible* crystalline structures are known to us.

But suppose the machine is now studying another kind of object: for instance, a hive built by wild bees. There it would obviously find all the signs indicating artificial origin: the simple and repeated geometrical structures of the honeycombs and the cells composing them, thanks to which the hive could be classified in the same category of objects as the Barbizon dwellings. What are we to make of this conclusion? We know the hive is 'artificial' insofar as it represents the product of the activity of bees. But we have good reasons for thinking that this activity is strictly automatic – immediate, but not consciously planned. At the same time, as good naturalists we view bees as 'natural' beings. Is there not a flagrant contradiction in considering the product of a 'natural' being's automatic activity as 'artificial'?

Carrying on the investigation, it would soon be seen that if there is contradiction, it results not from faulty programming but from the ambiguity of our judgments. For if the machine now inspects, not the hive, but the bees themselves, it can only see them as artificial, highly elaborated objects. The most superficial examination will reveal in the bee elements of simple symmetry: bilateral and translational. Above all, examining bee after bee, the computer will note that the extreme complexity of their structure (the number and position of abdominal hairs, for example, or the ribbing of the wings) is reproduced with extraordinary fidelity from one individual bee to the next. This is certain evidence that these creatures are the products of a deliberate, constructive, and highly sophisticated order of activity. On the basis of such conclusive documentation, the machine would certainly signal to the officials of the Martian NASA its discovery of an industry upon Earth beside which their own would probably seem primitive.

In this little excursion into near science fiction, our aim was only to illustrate the difficulty of defining the distinction, which might seem intuitively evident, between 'natural' and 'artificial' objects. In fact, on the basis of structural criteria, macroscopic ones, it is probably impossible to arrive at a definition of the artificial which, while including all 'veritable' artifacts, such as the products of human industry, would exclude objects so clearly natural as crystalline structures, and indeed, living beings themselves which we would equally like to classify among natural systems.

Looking for the cause of the confusion – or seeming confusion – to which the programme is leading, we may wonder whether it does not arise from our having wished to limit it to considerations only of form, of structure, of geometry, and so robbing our notion of an artificial object of its essential content: which is that any such object is defined or explained primarily by the function it has to fulfil, the performance its inventor expects of it. However, we shall soon find that by programming the machine so that henceforth it studies not only the structure but the eventual performance of the examined objects, we end up with still more disappointing results.

For supposing that this new programme does enable the machine correctly to analyse the structure and the performance of two series of objects such as horses running in a field and cars moving along a highway. The analysis would lead to the conclusion that these objects are closely comparable, each having a built-in capacity for swift movement, although over different surfaces, which accounts for their differences of structure. And if, to take another example, the machine were asked to compare the structure and performance of the eye of a vertebrate with that of a camera, the programme could not fail to recognize their close similarities: lenses, diaphragm,

objects endowed with a purpose

shutter, light-sensitive pigments: the same components could only have been introduced into both objects in order to obtain similar performances from them.

This example of functional adaptation in living beings is a classical one, and I have quoted it only to emphasize how arbitrary and pointless it would be to deny that the natural organ, the eye, represents the materialization of a 'purpose' – that of picking up images – while this is indisputably also the origin of the camera. It would be all the more absurd to deny it since, in the last analysis, the purpose which 'explains' the camera can only be the same as the one to which the eye owes its structure. Every artifact is a product made by a living being which through it expresses, in a particularly conspicuous manner, one of the fundamental characteristics common to all living beings without exception: that of being *objects endowed with a purpose or project*, which at the same time they show in their structure and execute through their performances (such as, for instance, the making of artifacts).

Rather than reject this idea (as certain biologists have tried to do) it must be recognized as essential to the very definition of living beings. We shall maintain that the latter are distinct from all other structures or systems present in the universe by this characteristic property, which we shall call *teleonomy*.

But it must be borne in mind that, while necessary to the definition of living beings, this condition is not sufficient, since it does not propose any objective criteria for distinguishing between living beings themselves and the artifacts produced by their activity.

It is not enough to point out that the project which gives rise to an artifact belongs to the animal that created it, and not to the artificial object itself. This obvious notion is still too subjective, as is proved by the difficulty of utilizing it in the computer programme: for how could the machine decide that the project of picking up images – the project represented by the camera – belongs to some object other than the camera itself? Simply by examination of the finished structure and by analysis of its performance it is possible to identify the project, but not its author or source.

To achieve this we must have a programme which studies not only the actual object but its origin, its history, and, for a start, the method of its construction. There is, in principle at least, no obstacle in the formulation of such a programme. Even if it were rather crudely compiled, it would enable us to discern a radical difference between any artifact, however highly perfected, and a living being. The machine could not fail to note that the macroscopic structure of an artifact (whether a honeycomb, a dam built by beavers, a paleolithic hatchet, or a spacecraft) results from the application to its constituent materials of forces *exterior* to the object itself. Once complete, this macroscopic structure is not evidence of inner forces of cohesion between atoms or molecules constituting its material (only conferring upon it its general properties of density, hardness, ductility, etc.), but of the *external* forces that have shaped it.

On the other hand, the programme will have to register the fact that the structure of a living being results from a totally different process, in that it owes almost nothing to the action of outside forces, but everything, *self-constructing machines* from its overall shape down to its tiniest detail, to 'morphogenetic' interactions within the object itself. It is thus a structure giving proof of an autonomous determinism: precise, rigorous, implying a virtually total 'freedom' with respect to external agents or conditions – which are capable, to be sure, of impeding this development, but not of directing it, nor of prescribing its organizational scheme to the living object. By the autonomous and spontaneous character of the morphogenetic processes that build the macroscopic structure of living beings, the latter are entirely distinct from artifacts, as they are, moreover, from the majority of natural objects whose macroscopic morphology largely results from the influence of external agents. There is one exception: that, once again, of crystals, whose characteristic geometry reflects microscopic interactions occurring within the object itself. By this criterion alone, crystals would have to be classified with living beings, while artifacts and natural objects, both fashioned by outside agents, would form another class.

That crystalline structures and the structures of living beings could be related by applying this criterion, after the criteria of regularity and repetition, might well give our programmer food for thought. Even if unversed in modern biology, he would wonder whether the internal forces which give living beings their macroscopic structure might be of the same nature as the microscopic interactions responsible for crystalline morphologies. That this is indeed the case is one of the main themes to be developed later in this essay; but for the moment we are trying by the most general criteria to define the macroscopic properties that differentiate living beings from all other objects in the universe.

Having 'discovered' that an internal, autonomous determinism guarantees the formation of the extremely complex structures of living beings, our programmer (with no training in biology, but an information specialist by profession) must necessarily see that such structures represent a considerable quantity of information whose source has still to be identified: for all information expressed – and hence received – presupposes a source.

Let us assume that, continuing his investigation, our programmer at last makes his final

self-reproducing
machines

discovery: that the source of the information expressed in the structure of a living being is *always* another, structurally identical, object. He has now identified the source and detected a third remarkable property in these objects: their ability to reproduce and to transmit *ne varietur* the information corresponding to their own structure; very valuable information, since it describes an organizational scheme which is exceedingly complex and is also preserved intact from one generation to the next. We will name this property *invariant reproduction*, or simply *invariance*.

By the property of invariant reproduction we again find living beings and crystalline structures related and unlike all other known objects in the universe. It is known that certain chemicals in supersaturated solution do not crystallize unless the solution has been inoculated with crystal seeds. We know too that in cases of a chemical capable of crystallizing into two different systems, the structure of the crystals appearing in the solution will be determined by that of the seed employed. Crystalline structures, however, represent a quantity of information inferior by several orders of magnitude to that transmitted from one generation to another in the simplest living beings we are acquainted with. By this criterion – purely quantitative, of course – living beings may be distinguished from all other objects, crystals included.

Let us now leave our Martian programmer to his thoughts. The only object of this imaginary experiment was to compel us to 'rediscover' the most general properties that characterize living beings and distinguish them from the rest of the universe. Let us now admit to a familiarity with modern biology, so as to analyse more closely and to define more precisely, if possible quantitatively, the properties in question. We have found three: teleonomy, autonomous morphogenesis, and reproductive invariance.

Of these, reproductive invariance is the easiest to define quantitatively. Since this is the capacity to reproduce highly ordered structure, and since a structure's degree of order can be defined in units of information, we shall say that the 'invariance content' of a given species is equal to the amount

strange properties: invariance and teleonomy

of information which, transmitted from one generation to the next, ensures the preservation of the specific structural standard. We shall see later on that with the help of a

few assumptions it is possible to reach an estimate of this amount of information.

This in turn will enable us to bring into better focus the notion most immediately and plainly inspired by the examination of the structures and performances of living beings, that of teleonomy. This when analysed nevertheless appears to be a profoundly ambiguous concept, since it implies the subjective idea of 'project'. Take the example of the camera: if we agree that this object's existence and structure realize the 'project' of capturing images, we must also agree that a similar project is accomplished with the emergence of the eye of a vertebrate.

But it is only as a part of a more comprehensive project that each individual project, whatever it may be, has any meaning. All the functional adaptations in living beings, like all the artifacts they produce, fulfil particular projects which may be seen as so many aspects or fragments of a unique primary project, which is the preservation and multiplication of the species.

To be more precise, we shall define the essential teleonomic project as consisting in the transmission from generation to generation of the invariance content characteristic of the species. All structures, performances and activities contributing to the success of the essential project will hence be called 'teleonomic'.

This allows us to put forward at least the *principle* of a definition of a species' 'teleonomic level'. All teleonomic structures and performances can be regarded as corresponding to a certain quantity of information which must be transmitted for these structures to be realized and these performances accomplished. Let us call this quantity 'teleonomic information'. A given species' 'teleonomic level' may then be said to correspond to the quantity of information which, on the average and per individual, must be transferred to assure the generation-to-generation transmission of the specific content of reproductive invariance.

It will be readily seen that, in this or that species higher or lower on the animal scale, the achievement of the fundamental teleonomic project (i.e., invariant reproduction) calls various more or less elaborate and complex structures and performances into play. It must be stressed that it is a matter not only of the activities directly linked with reproduction itself, but all those that contribute – however indirectly – to the species' survival and multiplication. In higher mammals, for example, the play of the young is an important element of psychic development and social integration. Therefore this activity has teleonomic value, in furthering the cohesion of the group, a condition for survival and for the expansion of the species. It is the degree of complexity of all these performances or structures, conceived as having the function of serving the teleonomic purpose, that we would like to estimate.

This magnitude, while theoretically definable, is not measurable in practice. Still, it may serve as a rule of thumb for ranking different species or groups upon a 'teleonomic scale'. As an extreme example, imagine a bashful poet who does not dare to declare his passion to the woman he loves and can only express it symbolically, in the poems he dedicates to her. Suppose that at last, conquered by these refined compliments, the lady surrenders to the poet's desire. His verses will have contributed to the success of his essential project, and the information they contain must therefore be counted in the sum of the teleonomic performances ensuring transmission of genetic invariance.

Clearly there is no analogous performance in the successful accomplishment of the project in other animal species, the mouse for instance. But – and this is the important point – the genetic invariance content is about the same in the mouse and the human being (and in all mammals, for that matter). *The two magnitudes we have been trying to define are therefore quite distinct.*

This leads us to consider a most important question concerning the relationship between the three properties we singled out as characteristic of living beings. The fact that the computer programme identified them successively and independently does not prove that they are not simply three manifestations of a single, more basic, more secret property, inaccessible to any direct observation. If this were so, the drawing of distinctions among the properties, the seeking of different definitions for them, might be merely delusion and arbitrariness. Far from shedding light on the real problem, far from tracking down 'the secret of life' and truly dissecting it, we would only be exorcizing it.

It is perfectly true that these three properties – teleonomy, autonomous morphogenesis, and reproductive invariance – are closely interconnected in all living beings. Genetic invariance expresses and reveals itself only through, and thanks to, the autonomous morphogenesis of the structure that constitutes the teleonomic apparatus.

It must first be observed that these three concepts do not all have the same standing. Whereas invariance and teleonomy are indeed characteristic 'properties' of living beings, spontaneous structuration ought rather to be considered a mechanism. Further on we shall see that this mechanism intervenes both in the elaboration of teleonomic structures and in the reproduction of invariant information. That it finally accounts for the latter two properties does not, however, imply that they should be regarded as one. It remains possible – it is in fact methodologically indispensable – to maintain a distinction between them, and this for several reasons:

1. One can at least *imagine* objects capable of invariant reproduction but devoid of any teleonomic apparatus. Crystalline structures offer one example of this, at a level of complexity admittedly very much lower than that of all known living organisms.

2. The distinction between teleonomy and invariance is more than a mere logical abstraction. It is warranted on grounds of chemistry. Of the two basic classes of biological macromolecules, one, that of proteins, is responsible for almost all teleonomic structures and performances; while genetic invariance is linked exclusively to the other class, that of nucleic acids.

3. Finally, as will be seen in the next chapter, this distinction is assumed, explicitly or otherwise, in all theories, all ideological constructions (religious, scientific, or philosophical) pertaining to the biosphere and to its relationship to the rest of the universe.

Living creatures are strange objects, as men of all past ages must have been more or less confusedly aware. The development of the natural sciences beginning in the seventeenth century, and their flowering from the nineteenth century on, instead of effacing this impression of strangeness made it even sharper. In observing the physical laws governing macroscopic systems, the very existence of living organisms seemed to constitute a paradox, violating certain of the fundamental principles of modern science. Which principles? It is not immediately clear. So what is needed is a precise analysis of the nature of this – or these – 'paradoxes'. This will give us occasion to specify the relative position, vis-à-vis physical laws, of the two essential properties that characterize living organisms: reproductive invariance and structural teleonomy.

Indeed at first glance invariance appears to constitute a profoundly paradoxical property, since the maintaining, the reproducing, the multiplying of highly ordered structures seems incompatible with the second law of thermodynamics. This law lays it down that no macroscopic system can evolve otherwise than in a downward direction, toward degradation of the order that characterizes it.

the 'paradox' of invariance

However, this prediction of the second law is valid, and verifiable, only if we are considering the overall evolution of an *energetically isolated* system. Within such a system, in one of its phases, we may see ordered structures take shape and grow without that system's overall evolution ceasing to comply with the second law. The best example of this is afforded by the crystallization of a saturated solution. The thermodynamics of such a system are well understood. The local increase of order, represented by the assembling of initially unordered molecules into a perfectly defined crystalline network, is 'paid for' by a transfer of thermal energy from the crystalline phase to the solution: the entropy – or disorder – of the system as a whole augments to the extent stipulated by the second law.

This example shows that, within an isolated system, a local increase of order is compatible with the second law. We have pointed out, however, that the degree of order represented by even the simplest organism is incomparably higher than that which a crystal defines. It must now be asked whether the conservation and invariant multiplication of such structures is also compatible with the second law. This can be verified through an experiment closely comparable with that of crystallization.

We take a millilitre of water having in it a few milligrams of a simple sugar, such as glucose, as well as some mineral salts containing the essential elements that enter into the chemical constituents of living organisms (nitrogen, phosphorus, sulphur, etc.). In this medium we grow a bacterium, for example *Escherichia coli* (length, 2 microns; weight, approximately 5×10^{-13} grams). Within thirty-six hours the solution will contain several thousand million bacteria. We shall find that about 40 per cent of the sugar has been converted into cellular constituents, while the remainder has been oxidized into carbon dioxide and water. By carrying out the entire experiment in a calorimeter, one can draw up the thermodynamic balance for the operation and determine that, as in the case of crystallization, the entropy of the system as a whole (bacteria plus medium) has increased a little more than the minimum prescribed by the second law. Thus, while the extremely complex system represented by the bacterial cell has not only been conserved but has multiplied several thousand million times, the thermodynamic debt corresponding to the operation has been duly settled.

No definable or measurable violation of the second law has occurred. Nonetheless, our physical intuition cannot fail to be deeply disturbed as we watch this phenomenon, whose strangeness is even more appreciable than before the experiment. Why? Because we see very clearly that this process is turned or oriented in one exclusive direction: the multiplication of cells. These it is true do not violate the laws of thermodynamics; on the contrary, they not only obey them, they use them, as a good engineer would, to carry out the project with maximum efficiency and bring about the 'dream' (as François Jacob has put it) of every cell: to become two cells.

Later we shall try to give an idea of the complexity, the subtlety and the efficiency of the chemical machinery necessary to the accomplishment of a project demanding the synthesis of several hundred different organic constituents; their assembly into several thousand macromolecular species;

and the mobilization and utilization, where necessary, of the chemical potential liberated by the oxidation of sugar, i.e., in

teleonomy and the principle of objectivity

the construction of cellular organelles. There is, however, no physical paradox in the invariant reproduction of these structures: invariance is bought at its exact thermodynamic price, thanks to the perfection of the teleonomic apparatus which, grudging of calories, in its infinitely complex task attains a level of efficiency rarely approached by man-made machines. This apparatus is entirely logical, wonderfully rational, and perfectly adapted to its purpose: to preserve and reproduce the structural norm. And it achieves this, not by transgressing physical laws, but by exploiting them to the exclusive advantage of its personal idiosyncrasy. It is the very existence of this purpose, at once both pursued and fulfilled by the teleonomic apparatus, that is the 'miracle'. Miracle? No, the real problem lies at another, deeper level than that of physical laws; it lies in our understanding, in our intuition of the phenomenon. There is really no paradox or miracle, but a flagrant *epistemological contradiction*.

The cornerstone of the scientific method is the postulate that nature is objective. In other words, the *systematic* denial that 'true' knowledge can be reached by interpreting phenomena in terms of final causes – that is to say, of 'purpose'. An exact date may be given for the discovery of this canon. The formulation by Galileo and Descartes of the principle of inertia laid the groundwork not only for mechanics but for the epistemology of modern science, by abolishing Aristotelian physics and cosmology. Certainly, neither reason, nor logic, nor observation, nor even the idea of their systematic confrontation had been lacking among Descartes's predecessors. But science as we understand it today could not have been built upon those foundations alone. It still needed the strict censorship implicit in the postulate of objectivity – pure, and impossible to demonstrate. For it is obviously impossible to imagine an experiment proving the *nonexistence* anywhere in nature of a purpose, of a pursued end.

But the postulate of objectivity is consubstantial with science, and has guided the whole of its prodigious development for three centuries. It is impossible to escape it, even provisionally or in a limited area, without departing from the domain of science itself.

Objectivity nevertheless obliges us to recognize the teleonomic character of living organisms, to admit that in their structure and performance they decide on and pursue a purpose. Here therefore, at least in appearance, lies a profound epistemological contradiction. In fact the central problem of biology lies with this very contradiction, which, if it is only apparent, must be resolved, or else proved to be radically insoluble, if that should turn out indeed to be the case.

2

Vitalisms and Animisms

As the teleonomic properties of living beings appear to challenge one of the basic postulates of the modern theory of knowledge, any philosophical, religious, or scientific view of the world must, *ipso facto*, offer an implicit if not an explicit solution to this problem. Every solution in its turn, whatever the motivation behind it, just as inevitably implies a hypothesis as to the causal and temporal precedence, in relation to each other, of the two properties characteristic of living beings: invariance and teleonomy.

the priority relationship between invariance and teleonomy: a fundamental dilemma

In a later chapter we will give an exposition of, and justifications for, the only hypothesis that modern science here deems acceptable: namely, that invariance necessarily precedes teleonomy. Or, to be more explicit, the Darwinian idea that the initial appearance, evolution, and continuous refining of ever more intensely teleonomic structures are due to disturbances occurring in a structure *which already possesses the property of invariance* – and hence is capable of preserving the effects of chance and thereby submitting them to the play of natural selection.

Of course the theory briefly and dogmatically sketched here is not that of Darwin himself, who in his day could not have had any idea of the chemical mechanisms of reproductive invariance, nor of the nature of the perturbations which affect these mechanisms. But it is no disparagement of Darwin's genius to note that the selective theory of evolution could not take on its full significance, precision, and certainty until less than twenty years ago.

The selective theory is the only one so far proposed that, while ranking teleonomy as a secondary property deriving from invariance – alone seen as primary – is consistent with the postulate of objectivity. It is also the only one not merely compatible with modern physics but based squarely upon it, without restrictions or additions. In short, the selective theory of evolution assures the epistemological coherence of biology and gives it its place among the sciences of 'objective nature'. This is a powerful argument in favour of the theory; but not convincing enough to justify it.

All other concepts put forward explicitly to solve the problem of the strangeness of living beings, or which are implicitly contained in religious ideologies and most of the great philosophical systems, assume the reverse hypothesis: that *invariance is safeguarded, ontogeny guided,* and *evolution oriented* by an initial teleonomic principle, of which all these phenomena are the purported manifestations. I shall devote the remainder of this chapter to a schematic analysis of the logic of these interpretations, very diverse in appearance but all implying the renunciation, partial or total, admitted or not, conscious or otherwise, of the postulate of objectivity. It will be convenient here to classify these concepts (rather arbitrarily, it is true) under one of two headings, according to the nature and supposed extension of the teleonomic principle they invoke.

Thus on one side we may place a first group of theories involving a teleonomic principle which operates only within the biosphere, in the heart of 'living matter'. These theories, which I shall call *vitalist,* therefore imply a radical distinction between living beings and the inanimate world.

And on the other side we may group together the concepts that posit a *universal* teleonomic principle, responsible for the course of affairs throughout the cosmos as well as within the biosphere, where this principle is expressed simply in a more precise and intense manner. These theories see living beings as the most highly elaborated, most perfect products of a universally oriented evolution, which has culminated, because it had to, in man and mankind. These concepts I shall call *animist*: they are in many respects more interesting than the vitalist theories, of which I shall give only a brief summary.[1]

Among vitalist theories a wide variety of tendencies may be discerned. Here we shall be content with distinguishing between what I shall refer to as 'metaphysical vitalism' and 'scientistic vitalism'.

The most illustrious proponent of a metaphysical vitalism was Henri Bergson. Thanks to an engaging style and a metaphorical dialectic bare of logic but not of poetry, his

*metaphysical
vitalism*

philosophy achieved immense success. Today it seems to have become almost completely discredited; but in my youth no one could hope to pass

his baccalaureate examination unless he had read *Creative Evolution*. This philosophy, as will be recalled, rests entirely upon a certain idea of life conceived as an *élan*, a 'current', absolutely distinct from inanimate matter but contending with it, 'traversing' it so as to force it into organized form. Contrary to almost all other vitalisms and animisms, that of Bergson is not finalist: it refuses to imprison life's essential spontaneity in any kind of predetermination. Evolution, identified with the *élan vital* itself, can therefore have neither final nor efficient causes. Man is the supreme stage at which evolution has arrived, without having sought or foreseen it. He is rather the sign and proof of the total freedom of the creative *élan*.

1. It may be well to stress that I am here employing the qualifying 'animist' and 'vitalist' in a special sense, somewhat different from current usage.

This conception is linked with another, considered fundamental by Bergson: rational intelligence is an instrument of knowledge specially designed for mastering inert matter but totally incapable of apprehending life's phenomena. Only instinct, consubstantial with the *élan vital*, can give a direct, global insight into them. Every analytical and rational statement about life is therefore meaningless, or rather irrelevant. The high development of rational intelligence in *Homo sapiens* has led to a serious and regrettable impoverishment of his powers of intuition, a lost treasure which we today must strive to recover.

I shall not try to discuss this philosophy (which indeed does not lend itself to discussion). A captive of logic, and weak in global intuitions, I feel I am not qualified. However, I do not regard Bergson's attitude as unimportant; quite the contrary. Conscious or unconscious rebellion against the rational, respect given to the *id* at the expense of the *ego*, are hallmarks of our times, and so is creative spontaneity. Had Bergson written in a less limpid language, a more 'profound' style, he would be read again today.[2]

There have been many 'scientific' vitalists, and they include some very distinguished scholars. But while fifty years ago the vitalists were recruited from among biologists (of whom the most renowned, Driesch, gave up embryology for philosophy), those of our day, like Professors Elsässer and Polanyi, come mainly from the physical sciences. It is understandable, certainly, that physicists

scientistic vitalism

2. Bergson's thought, it need hardly be said, is not lacking in obscurity or patent contradictions. One may well question, for example, whether Bergsonian dualism is essential: should it not perhaps be seen as deriving from a more basic monism? (C. Blanchard, in a personal communication.) My intention here is, of course, not to explore Bergson's thought in its ramification, but only in those implications which most directly concern the theory of living systems.

should be even more impressed than biologists by the strangeness of living things. Here, for example, briefly summarized, is Elsässer's position.

The strange properties, invariance and teleonomy, are doubtless not fundamentally opposed to physics; but the physical forces and chemical interactions brought to light by the study of nonliving systems *do not fully account for these properties*. Hence it must be accepted that certain principles, which become *added* to those of physics, are operative in living matter, but not in nonliving systems where, consequently, these electively vital principles could not be discovered. It is these principles – or, to borrow from Elsässer's terminology, these 'biotonic laws' – that must be elucidated.

Even the great Nils Bohr himself, it seems, did not dismiss such hypotheses. But he did not claim to have proof that they were necessary. Are they? In the end, this is the whole question. That is what Elsässer and Polanyi assert. The least one can say is that the arguments of these physicists are oddly weak and woolly.

These arguments concern respectively each of the strange properties. As to invariance, its mechanism is well enough known today for us to affirm that no nonphysical principle is required for its interpretation.[3]

This leaves teleonomy or, more exactly, the morphogenetic mechanisms which put teleonomic structures together. It is perfectly true that embryonic development is in appearance one of the most miraculous phenomena in the whole of biology. It is also true that these phenomena, admirably described by embryologists, still largely (for technical reasons) elude genetic and biochemical analysis, which alone could lead to an understanding of them. The attitude of the vitalists who feel that physical laws are – or will prove to be – insufficient to explain embryogenesis draws its justification,

3. See Chapter 6.

therefore, not from precise knowledge or from definite observations, but only from our present-day ignorance.

On the other hand, our understanding of the molecular control mechanisms that regulate cellular growth and activity has progressed considerably and will no doubt soon contribute to the interpretation of organic development. We shall come to a discussion of these mechanisms in Chapter 4, and shall then have more to say about certain vitalist arguments. For vitalism to survive, it needs the survival in biology, if not of actual paradoxes, at least of a few 'mysteries'. Developments in molecular biology over the past two decades have singularly narrowed the domain of the mysterious, leaving little open to vitalist speculation but the field of subjectivity: that of consciousness itself. There is no great risk in predicting that also in this area, for the time being still 'reserved', such speculation will prove as sterile as in all the others where it has been practised up to now.

Animist conceptions, as I have already said, are in many respects a great deal more interesting than vitalist ideas. Reaching back to mankind's infancy, perhaps to before the appearance of *Homo sapiens*, they are still deep-rooted in the soul of modern man.

Our ancestors were probably only very dimly aware of the strangeness of their condition. They did not have the reasons we have today for feeling themselves strangers in the universe

the animist projection and the 'ancient covenant'

upon which they opened their eyes. What did they see first? Animals, plants; beings whose nature they could at once guess to be similar to their own. Plants grow, seek sunlight, die; animals stalk their prey, attack their enemies, feed and protect their young; males fight for the possession of a female. Plants and animals like man himself were easy to explain: they all have a purpose: to live and to go on living in their progeny, even at the price of death. Its purpose explains the being, and the being makes sense only through the purpose animating it.

But around them our ancestors also saw other objects, far more mysterious: rocks, rivers, mountains, the thunderstorm, the rain, the stars in the sky. If these exist it must also be for a purpose, to nourish which they had also to have a spirit or soul. For those early human beings the world's strangeness was resolved thus: no inanimate objects really exist, for such a thing would be incomprehensible. In the river's depths, on the mountaintop, more subtle spirits pursue vaster and more impenetrable designs than the visible ones animating men and beasts. Thus, in nature's forms and events our forebears saw the action of forces either benign or hostile, but never indifferent – never totally alien.

Animist belief, as I am visualizing it here consists essentially in a projection into inanimate nature of man's awareness of the intensely teleonomic functioning of his own central nervous system. It is, in other words, the hypothesis that natural phenomena can and must be explained in the same manner, by the same 'laws', as subjective human activity, conscious and purposive. Primitive animism formulated this hypothesis with complete candour, frankness, and precision, populating nature with gracious or awe-inspiring myths and myth-figures which have for centuries nourished art and poetry.

It would be wrong to smile, even with the affection and respect aroused by childhood. Do we imagine that modern culture has really given up the subjective interpretation of nature? Animism established a covenant between nature and man, a profound alliance outside which there seems to stretch only terrifying solitude. Must we break this tie because the postulate of objectivity requires it? Ever since the seventeenth century the history of ideas testifies to the profuse efforts of the greatest minds to avert that break, to renew the 'ancient covenant'. Think of such mighty endeavours as those of Leibniz, or of Hegel's colossal and ponderous monument. But idealism has not by any means been the only refuge for a cosmic animism. At the very core of certain ideologies claiming to be founded upon science, the animist projection, in a more or less disguised form, turns up again.

The biological philosophy of Teilhard de Chardin would not merit attention but for the startling success it has encountered even in scientific circles. A success which tells of the eagerness, of the need to revive *scientistic* the covenant. Teilhard revives it, *progressism* without disguise. His philosophy, like Bergson's, is based entirely upon an initial evolutionist postulate. But, unlike Bergson, he has the evolutive force operating throughout the entire universe, from elementary particles to galaxies: there is no 'inert' matter, and therefore no essential distinction between 'matter' and 'life'. His wish to present this concept as 'scientific' leads Teilhard to base it upon a new definition of energy. This is somehow distributed between two vectors, one of which would (I presume) be 'ordinary' energy, whereas the other would correspond to the upward evolutionary surge. The biosphere and man are the latest products of this ascent along the spiritual vector of energy. This evolution is to continue until all energy has become concentrated along the spiritual vector: that will be 'point omega'.

Although Teilhard's logic is hazy and his style laborious, some who do not entirely accept his ideology yet allow it a certain poetic grandeur. Personally, I am shocked by the intellectual spinelessness of this philosophy. It appears to me above all to show a systematic complacency, a willingness to conciliate at any price, to come to any compromise. Perhaps Teilhard was not in vain a member of that order which, three centuries earlier, Pascal assailed for its theological laxness.

It was of course not Teilhard who discovered the idea of

re-establishing the old animist covenant with nature, or of founding a new one through a universal theory according to which the evolution of the biosphere culminating in man would be part of the smooth onward flow of cosmic evolution itself. This idea was in fact the central theme of nineteenth-century scientistic progressism. One finds it at the very heart of Spencer's positivism and of Marx and Engels' dialectical materialism. The unknown and *unknowable* force which, according to Spencer, operates throughout the universe, creating variety, coherence, specialization and order, plays what amounts to exactly the same role in Teilhard's 'ascending' energy: human history is the extension of biological evolution, itself a component part of cosmic evolution. Thanks to this single principle, man at last finds his eminent and necessary place in the universe, along with certainty of the progress to which he is pledged.

Spencer's differentiating force, like Teilhard's ascending energy, is a clear instance of animist projection. In order to give meaning to nature, to ensure that man is not separated from it by a bottomless pit, and, finally, to make it decipherable and intelligible, *a purpose had to be restored to it*. Failing a soul to harbour this purpose, one injects into nature an evolutive, an ascending, 'force', which in effect means abandoning the postulate of objectivity.

Among the scientistic ideologies of the nineteenth century the most powerful one, which still today wields a profound influence far beyond the already vast circle of its followers, is of course Marxism. Hence it is most revealing to note that, in their wish to base their social doctrines on nature's own laws, Marx and Engels too resorted, more clearly and deliberately than Spencer, to 'animist projection'.

the animist projection in dialectical materialism

Indeed I do not see how else one can interpret the famous 'inversion' by which Marx substitutes dialectical materialism for the idealist dialectic of Hegel.

Hegel's postulate, that the most general laws governing the universe in its evolution are of a dialectical order, has its proper place within a system which admits no permanent and authentic reality except mind. If all events, all phenomena are but partial manifestations of 'an idea that thinks itself', it is legitimate to look for the most immediate expression of the universal laws in our subjective experience of the thinking process. And since thought proceeds dialectically, 'the laws of dialectic' govern the whole of nature. But to retain these subjective laws just as they are and make them into those of a purely material universe is to effect the animist projection quite openly and with all its consequences beginning with the scrapping of the postulate of objectivity.

Neither Marx nor Engels made a detailed analysis to justify the logic of this inversion of dialectics. However, using the numerous examples of its application, given particularly by Engels in his *Anti-Dühring* and *Dialectics of Nature*, one can attempt to reconstruct the underlying thought of the founders of dialectical materialism. Its essential tenets would be these:

1. Movement is the mode of existence of matter.

2. The universe, defined as the totality of matter, which alone exists, is in a state of perpetual evolution.

3. True knowledge of the universe is that which contributes to the understanding of this evolution.

4. But this knowledge is obtained only in the interaction, itself evolutive and a cause of evolution, between man and matter (or, more exactly, the 'rest' of matter). All true knowledge is therefore 'practice'.

5. Consciousness pertains to this cognitive interaction. Conscious thought consequently reflects the movement of the universe itself.

6. Since, then, thought is a part and reflection of universal movement, and since its movement is dialectical, the evolutionary law of the universe itself must be dialectical. This explains and justifies the use of such terms as *contradiction*, *affirmation*, and *negation* in connection with natural phenomena.

7. The dialectic is constructive (thanks notably to the 'third law'). Therefore the evolution of the universe is itself ascendant and constructive. Its highest expression is human society, consciousness, thought, all necessary products of this evolution.

8. Through its emphasis on the evolutionary essence of the structures of the universe, dialectical materialism goes far beyond the materialism of the eighteenth century which, founded upon classical logic, was limited to recognizing only mechanical interactions between supposedly invariant objects, and therefore remained incapable of evolutionary thinking.

One may of course criticize this reconstruction and deny that it reflects Marx and Engels' authentic thought. But that is really only secondary. The influence of an ideology depends upon the meaning it maintains in the minds of its adepts, and which is spread by later commentators. Countless texts show that the above summary is a legitimate representation of at least the 'vulgate' of dialectical materialism. One example will do, especially significant because its author, J. B. S. Haldane, was an eminent modern biologist. He wrote in his preface to the English translation of *Dialectics of Nature*:

Marxism has a twofold bearing on science. In the first place Marxists study science among other human activities. They show how the scientific activities of any society depend on its changing needs, and so in the long run on its productive methods, and how science changes the productive methods, and therefore the whole society. . . . But secondly Marx and Engels were not content to analyse the changes in society. In dialectics they saw the science of the general laws of change, not only in society and in human thought, but in the external world which is mirrored by human thought. That is to say it can be applied to problems of 'pure' science as well as to the social relations of science.[4]

The external world 'mirrored by human thought': that indeed sums it up. The logic of the inversion obviously requires that this mirroring be more than a fairly faithful transposition of the external world. For dialectical materialism it is indispensable that the *Ding an sich* – the thing or phenomenon in itself – should reach the level of consciousness unaltered and undiminished, with none of its properties suppressed by selection. The external world must be literally presented to consciousness in the full integrity of its structures and movement.[5]

No doubt certain of Marx's own writings could be quoted in opposition to this concept. Nonetheless, it remains indispensable to the logical coherence of dialectical materialism, as later Marxists, if not Marx and Engels themselves, well realized. Let us not forget, moreover, that dialectical materialism was a relatively late addition to the socio-

4. Friedrich Engels, *Dialectics of Nature* (London, Lawrence and Wishart, 1940), p. vii.

5. From Henri Lefebvre (*Le Matérialisme dialectique* (Paris, PUF, 1949), p. 92) we take the following passage: 'Far more than a mere process of thought, dialectic exists prior to mind, inheres in being. It obtrudes itself upon mind. First we analyse the simplest stirring of thought; of the most abstract, the barest thought. In so doing we discover the most general categories and their concatenation. These we must next connect to the concrete movement, to the *given content*; we are then made aware of the fact that the process that involves the content and the self clarifies itself for us in the workings of the laws of dialectic. Contradictions in thought come not from thinking alone, from its weaknesses or incoherence; they come also from the content. Their interlocking tends toward the expression of the *total movement of the content* and lifts it to the level of consciousness and of reflection.'

economic edifice which Marx had already raised and one clearly intended to turn historical materialism into a 'science' based on the laws of nature itself.

Their insistence upon the 'perfect mirror' explains the dialectical materialists' dogged repudiation of any kind of critical epistemology, immediately condemned as 'idealist' or 'Kantian'. This attitude is certainly to some extent understandable in men living in the nineteenth century, contemporary witnesses to the first great scientific upheaval. It might then very well look as though, thanks to science, man was in the process of achieving direct mastery over nature, appropriating its very substance. Nobody doubted, for instance, that gravitation was one of the laws of nature itself, probed to its innermost depths.

the need for a critical epistemology

As we know, it was by a return to the very sources of knowledge itself that the spadework was done for the second age of science, that of the twentieth century. By the close of the nineteenth century the need for a critical epistemology once more became evident as a basic condition for the objectivity of knowledge. This critique was the concern not only of philosophers but also of scientists, who were led to incorporate it in the texture of theory itself. Only thus could the theory of relativity and quantum mechanics be developed.

Moreover, advances in neurophysiology and in experimental psychology are now beginning to disclose at least some aspects of the functioning of the nervous system. Enough to make it clear that the information the central nervous system furnishes to consciousness is, and probably can only be, in codified form, transposed, framed within pre-existing norms: in other words, assimilated and not just restored.

The thesis of pure reflection, of the perfect mirror not even inverting the image, seems to us more indefensible than ever. But it was really not necessary to wait for twentieth-century

the epistemological bankruptcy of dialectical materialism

scientific developments, to see the confusions and nonsense to which this notion was bound to lead. Poor Herr Dühring, an early recalcitrant, was given numerous examples of the dialectical interpretation of natural phenomena by Engels himself. There is the memorable one of the grains of barley, given as an illustration of the 'third law'.

> Let us take a grain of barley . . . if such a grain of barley meets with conditions which for it are normal, if it falls on suitable soil, then under the influence of heat and moisture a specific change takes place, it germinates; the grain as such ceases to exist, it is negated, and in its place appears the plant which has arisen from it, the negation of the grain. But what is the normal life-process of this plant? It grows, flowers, is fertilized and finally once more produces grains of barley, and as soon as these have ripened the stalk dies, is in its turn negated. As a result of this negation of the negation we have once again the original grain of barley, but not as a single unit, but ten, twenty or thirty fold. . . .

> It is the same [Engels adds a little further on] in mathematics. Let us take any algebraical magnitude whatever: for example, *a*. If this is negated, we get $-a$ (minus *a*). If we negate that negation, by multiplying $-a$ by $-a$, we get $+a^2$, i.e., the original positive magnitude, but at a higher degree, raised to its second power.[6]

And so forth.

These examples illustrate the scope of the epistemological

6. Friedrich Engels, *Herr Eugen Dühring's Revolution in Science (Anti-Dühring)* (London, Lawrence, 1935), pp. 154, 155.

disaster that follows the 'scientific' use of dialectical interpretations. Modern dialectical materialists ordinarily manage to avoid such absurdities. But to make dialectical contradiction the 'fundamental law' of all movement, all evolution, is still an attempt to systematize a subjective interpretation of nature, showing it to have an ascending, constructive, creative intent, a purpose; in short, to make nature decipherable and morally meaningful. This is 'animist projection' again, always recognizable whatever its disguises.

This interpretation is not only foreign to science but incompatible with it – and it has appeared as such every time the dialectical materialists, emerging from purely 'theoretical' verbiage, have tried to illuminate the path of experimental science by the light of their ideas. Although he had a sound knowledge of the science of his day, Engels himself was led, in the name of dialectics, to reject two of the greatest discoveries of the age: the second law of thermodynamics and (notwithstanding his admiration for Darwin) the theory of natural selection. It was by virtue of these same principles that Lenin assailed the epistemology of Mach; that, later, Zhdanov ordered Russian thinkers to scourge the Copenhagen school for 'its devilish Kantian mischief'; that Lysenko accused geneticists of maintaining a theory radically opposed to dialectical materialism, and therefore necessarily false. Despite the disclaimers of the Soviet geneticists, Lysenko was perfectly right: the theory of the gene as the hereditary determinant, invariant from generation to generation and even through hybridizations, is indeed completely irreconcilable with dialectical principles. It is by definition an idealist theory, since it is based on a postulate of invariance. The fact that we now know the structure of the gene and the mechanism of its invariant reproduction does not solve anything, for their description in modern biology is purely mechanistic. And so, at best, they are concepts ascribable to 'vulgar materialism', mechanistic, and hence 'objectively idealist', as

M. Althusser pointed out in his severe commentary upon my inaugural lecture before the Collège de France.

I have reviewed these various ideologies or theories briefly and very incompletely. Some may find that I have given a distorted, because partial, picture of them. I shall try to excuse myself by reminding the reader *the anthropocentric* that I have only tried here to point *illusion* out what these concepts hold, or imply, with respect to biology, and more especially the relationship they assume between invariance and teleonomy. It has been seen that, without exception, all take an initial teleonomic principle as the '*primum movens*' of evolution, whether of the biosphere alone or of the entire universe. In the eyes of modern scientific theory all these concepts are erroneous, not only for reasons of method (since in one way or another they imply abandonment of the postulate of objectivity) but for factual reasons, to be discussed in Chapter 6.

At the source of these errors, of course, is the anthropocentric illusion. The heliocentric theory, the concept of inertia, and the principle of objectivity were never enough to dissipate that ancient mirage. Far from dispelling the illusion, the theory of evolution at first seemed to endow it with a new reality by making man no longer the centre of the entire universe but its natural heir, awaited from time immemorial. At last God could die, replaced by this new and grandiose fantasy. The ultimate aim of science from now on would be to formulate a unified theory which, based on a small number of principles, would account for the whole of reality, biosphere and man included. This exalting certainty was the fare on which nineteenth-century scientistic progressism fed; it was a unified theory which, indeed, the dialectical materialists believed they had already formulated.

Because he saw it as jeopardizing the certainty that man

and human thought are necessary end-products of a cosmic progress, Engels was led to deny the second law. It is significant that he did so in the introduction to his *Dialectics of Nature*, and that he moved directly from this subject to an impassioned cosmological sermon in which he promised eternal recurrence, if not to the human species, at any rate to the 'thinking mind'. A recurrence indeed, but to one of mankind's most ancient myths.[7]

It was not until the second half of this century that the new anthropocentric illusion, propped on the theory of evolution, collapsed in its turn. I believe we can say today that a universal theory, however successful in other domains, could never encompass the biosphere, its structure, and its evolution as phenomena *deducible* from first principles.

the biosphere
a unique occurrence
nondeducible
from first principles

This proposition may appear obscure. Let us try to make it clearer. A universal theory would obviously have to extend to include relativity, the theory of quanta, and a theory of elementary particles. Provided certain initial conditions could be formulated, it would also contain a cosmology which

7. 'Hence,' Engels declares, 'we arrive at the conclusion that in some way, which it will later be the task of scientific research to demonstrate, the heat radiated into space must be able to become transformed into another form of motion, in which it can once more be stored up and rendered active. Thereby the chief difficulty in the way of the reconversion of extinct suns into incandescent vapour disappears. . . .

'But however often, and however relentlessly, this cycle is completed in time and space, however many millions of suns and earths may arise and pass away, however long it may last before the conditions for organic life develop, however innumerable the organic beings that have to arise and to pass away before animals with a brain capable of thought are developed from their midst, and for a short span of time find conditions suitable for life, only to be exterminated later without mercy, we have the certainty that matter remains eternally the same in all its transformations, that none of its attributes can ever be lost, and therefore, also, that with the same iron necessity that it will exterminate on earth its highest creation, the thinking mind, it must somewhere else and at another time again produce it' (*Dialectics of Nature*, pp. 23–5).

would forecast the general evolution of the universe. We know however (contrary to what Laplace believed, and after him the science and 'materialist' philosophy of the nineteenth century) that these predictions could be no more than statistical. The theory might very well contain the periodic table of elements, but could only determine the probability of existence of each of them. Likewise it would anticipate the appearance of such objects as galaxies or planetary systems, but would not in any case deduce from its principles the necessary existence of this or that object, event, or individual phenomenon – whether it be the Andromeda nebula, the planet Venus, Mount Everest, or last night's thunderstorm.

In a general manner the theory would anticipate the existence, the properties, the interrelations of certain *classes* of objects or events, but would obviously not be able to foresee the existence or the distinctive characteristics of any *particular* object or event.

The thesis I shall present in this book is that the biosphere does not contain a predictable class of objects or of events but is a particular event, certainly compatible indeed with first principles, but not *deducible* from those principles and therefore essentially unpredictable.

Let there be no misunderstanding. In saying that, as a class, living beings are not predictable upon the basis of first principles, I by no means intend to suggest that they are not *explicable* through these principles – that they transcend them in some way, and that other principles, applicable to living systems alone, must be invoked. In my view the biosphere is unpredictable for exactly the same reason – neither more nor less – that the particular configuration of atoms constituting this pebble I have in my hand is unpredictable. No one will blame a universal theory for not affirming and foreseeing the existence of this particular configuration of atoms; it is enough for us that this actual object, unique and real, be *compatible* with the theory. This object, according to the theory, is under no obligation to exist; but it has the right to.

That is enough for us as far as the pebble is concerned, but not as far as we ourselves are concerned. We would like to think ourselves necessary, inevitable, ordained from all eternity. All religions, nearly all philosophies, and even a part of science testify to the unwearying, heroic effort of mankind desperately denying its own contingency.

3

Maxwell's Demons

The concept of teleonomy implies the idea of an *oriented*, *coherent*, and *constructive* activity. By these standards proteins must be considered the essential molecular agents of teleonomic performance in living beings.

proteins as
molecular agents of
structural
and functional
teleonomy

1. Living beings are chemical machines. The growth and multiplication of all organisms require the accomplishing of thousands of chemical reactions whereby the essential constituents of cells are elaborated.

This is what is called 'metabolism'. It is organized along a great number of divergent, convergent, or cyclical 'pathways', each comprising a sequence of reactions. The precise adjustment and high efficiency of this enormous microscopic chemical activity are maintained by a certain class of proteins, the enzymes, playing the role of specific catalysts.

2. Like a machine, every organism, down to the very 'simplest', constitutes a coherent and integrated functional unit. Clearly enough, the functional coherence of so complex a chemical machine, which is autonomous as well, calls for a cybernetic system governing and controlling the chemical activity at numerous points. Especially as regards the higher organisms, we are still a long way from elucidating the entire structure of these systems. Nevertheless a great many of its elements are now known, and in all these cases it turns out that its essential agents are so-called 'regulatory' proteins which act, in effect, as detectors of chemical signals.

3. The organism is a self-constructing machine. Its macroscopic structure is not imposed upon it by outside forces. It shapes itself autonomously by dint of constructive internal interactions. Although our understanding of the mechanisms of development is still very imperfect, we can from now on state that the constructive interactions are microscopic and molecular, and that the molecules involved are essentially if not uniquely proteins.

Hence they are proteins which channel the activity of the chemical machine, ensure its coherent functioning, and put it together. All these teleonomic performances rest, in the final analysis, upon the proteins' so-called 'stereospecific' properties, that is to say upon their ability to 'recognize' other molecules (including other proteins) by their *shape*, this shape being determined by their molecular structure. There is here, quite literally, a microscopic discriminative (if not 'cognitive') faculty. We may say that any teleonomic performance or structure in a living being – whatever it may be – can, in principle at least, be analysed in terms of stereospecific interactions involving one, several, or a very large number of proteins.[1]

It is on the structure, on the shape of a given protein that the particular stereospecific discrimination constituting its function depends. To the extent that we could retrace the origin and evolution of this structure we would also be describing the origin and evolution of the teleonomic performance it discharges.

In the present chapter we shall discuss the specific catalytic function of proteins; in the next, their regulatory function;

1. I have deliberately oversimplified here. Certain DNA structures play a role that must be considered teleonomic. And certain RNAs (ribonucleic acids) constitute essential elements of the machinery which translates the genetic code (cf. Appendix 3, p. 181). However, particular proteins are also involved in these mechanisms which, at nearly every stage, bring interactions between proteins and nucleic acids into play. Discussion of these mechanisms may be omitted without affecting the analysis of teleonomic molecular interactions and their general interpretation.

and in Chapter 5 their constructive function. The problem of the origin of functional structures will be taken up in this last chapter and further discussed in the next.

One may indeed study the functional properties of a protein without having to refer to the details of its particular structure. (Actually, we have at present a thoroughly detailed knowledge of the three-dimensional structure of only some fifteen proteins.) A few general facts need recalling, however.

Proteins are very large molecules, ranging in molecular weight from 10,000 to 1,000,000 or more. These macromolecules are constituted by the sequential polymerization of components whose molecular weight is about 100, and which belong to the class of amino acids. Every protein thus contains from 100 to 10,000 amino acid residues. These very numerous residues belong, however, to only twenty different chemical species,[2] which are encountered in all living beings, from bacteria to man. This sameness of composition is one of the most striking illustrations of the fact that the prodigious diversity among *macroscopic* structures of living beings is based in fact on a profound and no less remarkable unity of *microscopic* structures. We shall have more to say about this.

On the basis of their general shape, one may divide proteins into two main classes:

a. The so-called 'fibrous' proteins are very elongated molecules which in living beings play a principally mechanical role, like that of the rigging on a sailing vessel; although the properties of some of these proteins (those found in muscle) are of great interest, we will not discuss them here.

b. The so-called 'globular' proteins are by far the most numerous and, through their functions, the most interesting; in these proteins the strands formed by the sequential polymerization of amino acids fold upon themselves in an exceedingly complex manner, thereby giving these molecules a compact, pseudo-globular shape.

2. See Appendix 1, pp. 172-5.

Even the simplest organisms contain a very great number of different proteins. It may be put at 2500 ± 500 for the bacterium *Escherichia coli* (weighing 5×10^{-13} grams and 2 microns in length, approximately). For the higher mammals such as man, one may suggest a figure of about a million.

Each of the thousands of chemical reactions that contribute to the development and performances of an organism is provoked electively by a particular enzyme-protein. One may say, scarcely oversimplifying, that in *the enzyme-proteins* the organism each enzyme exerts its *as specific catalysts* catalytic activity at only one point in the metabolism. It is above all through the extraordinary *electivity* of action they display that enzymes differ from nonbiological catalysts used in the laboratory or in industry. Some of the latter are exceedingly active – capable, in very slight quantity, of greatly accelerating various reactions. However, not one of these catalysts comes near the meanest ordinary enzyme for specificity of action.

This specificity is twofold:

1. Each enzyme catalyses only one type of reaction.

2. Among the sometimes very numerous compounds in the organism susceptible of undergoing that type of reaction, the enzyme, as a general rule, is active in regard to only one.

A few examples will help clarify these propositions. There is an enzyme, called fumarase, which catalyses the hydration of fumaric acid into malic acid. This reaction, illustrated

$$
\begin{array}{ccc}
\text{COOH} & & \text{COOH} \\
| & & | \\
\text{CH} & \xrightarrow{(+ H_2O)} & \text{HCOH} \\
\| & \rightleftharpoons & | \\
\text{CH} & & \text{CH}_2 \\
| & & | \\
\text{COOH} & & \text{COOH} \\
\text{(fumaric acid)} & & \text{(malic acid)}
\end{array}
$$

in the diagram opposite, is reversible, and the same enzyme also catalyses the dehydration of malic acid into fumaric acid.

Meanwhile there exists a geometric isomer of fumaric acid, maleic acid –

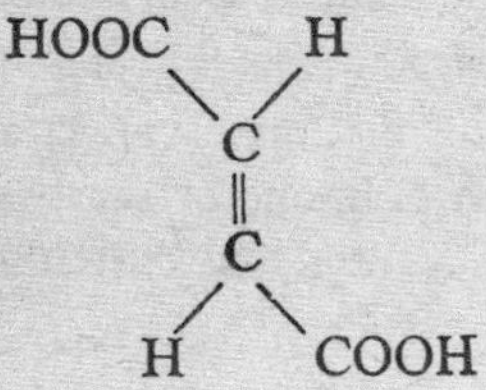

(fumaric acid)

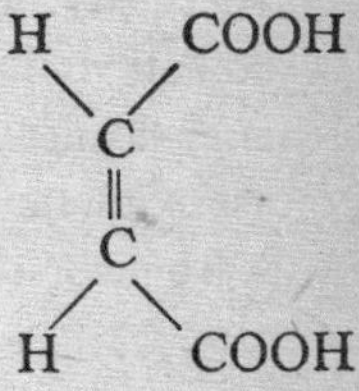

(maleic acid)

chemically capable of undergoing the same hydration. The enzyme is totally inactive with regard to the second.

But there are also two *optical* isomers of malic acid, which possesses an asymmetric carbon:[3]

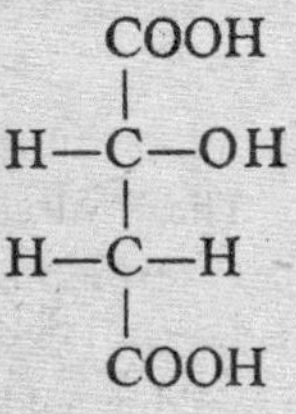

(L-malic acid)

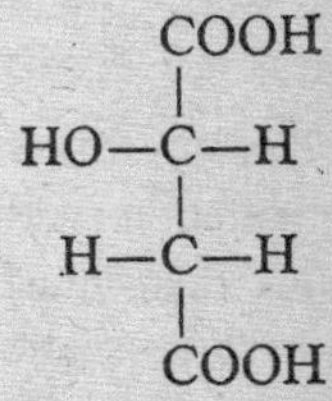

(D-malic acid)

Mirror-images of each other, these two compounds are chemically equivalent and practically inseparable by classical chemical techniques. Between the two the enzyme nevertheless exercises an absolute discrimination. Thus,

a. The enzyme dehydrates L-malic acid exclusively in order to produce fumaric acid exclusively; and

3. Compounds consisting of a carbon atom linked to four different groupings are thereby deprived of symmetry. They are said to be 'optically active', for the passage of polarized light through such compounds imparts to the plane of polarization a rotation to the left (L: levorotatory compounds) or to the right (D: dextrorotatory compounds).

b. Starting with fumaric acid, the enzyme produces L-malic acid exclusively, but not D-malic acid.

The rigorous distinction made by the enzyme between optical isomers is more than just a striking illustration of the *steric* specificity of enzymes. To begin with, one finds here the explanation for the previously mysterious fact that among the many chemical cellular constituents that are asymmetric (the case with the majority of them) only one of the two optical isomers is, as a general rule, represented in the biosphere.

But, in the second place, according to Curie's very general principle governing the conservation of symmetry, the fact that from an optically symmetrical compound (fumaric acid) an asymmetrical compound is obtained imposes the conditions that

a. The enzyme itself constitutes the 'source' of the asymmetry; hence, it must itself be optically active, which indeed it is;

b. The substrate's initial symmetry is lost in the course of its interaction with the enzyme-protein. The hydration reaction, then, must occur within a 'complex' formed by a temporary association between enzyme and substrate; in such a complex the initial symmetry of fumaric acid would be effectively lost.

'Stereospecific complex' as accounting for the specificity as well as for the catalytic activity of enzymes – the concept is of key importance. We shall return to it after discussing some further examples.

In certain bacteria another enzyme exists, called aspartase, which also acts upon fumaric acid alone, to the exclusion of every other compound and notably of its geometric isomer, maleic acid. The reaction of 'addition upon a double bond' catalysed by this enzyme is a close analogy to the preceding one. This time it is not a molecule of water but of ammonia that is condensed with fumaric acid to give an amino acid, aspartic acid:

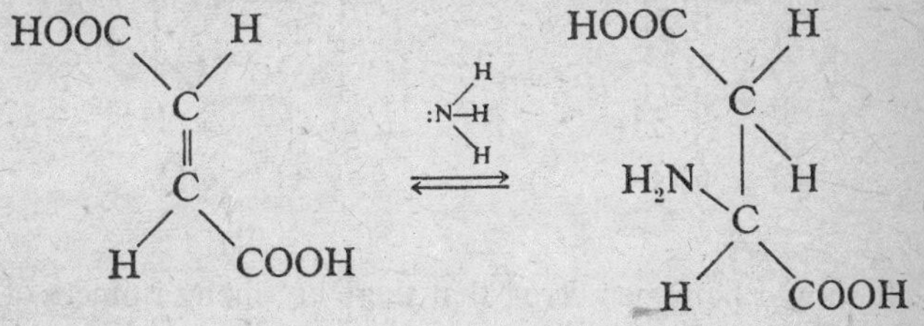

(fumaric acid)   (L-aspartic acid)

Aspartic acid possesses an asymmetrical carbon atom; it is therefore optically active. As in the preceding case, the enzymic reaction produces exclusively one of the isomers, the one of the L series, called a 'natural' isomer because amino acids entering into the composition of proteins all belong to the L series.

The two enzymes aspartase and fumarase thus discriminate strictly, not only between the optical and geometrical isomers of their substrates and products, but likewise between the molecules of water and ammonia. One is led to say that these latter molecules also enter into the composition of the stereo-specific complex, within whose framework the addition reaction is produced; and that in this complex the molecules are rigorously positioned in respect to each other. Both the specificity of action and the stereospecificity of the reaction would seem to result from this positioning.

From the preceding examples the existence of a stereo-specific complex, as intermediary in enzymic reactions, could be deduced only as an explanatory hypothesis. But in certain favourable cases the existence of this complex may be demonstrated directly. One such case is that of the enzyme called β-galactosidase, which specifically catalyses the hydro-lysis of compounds possessing the structure labelled (A) in the diagram overleaf.

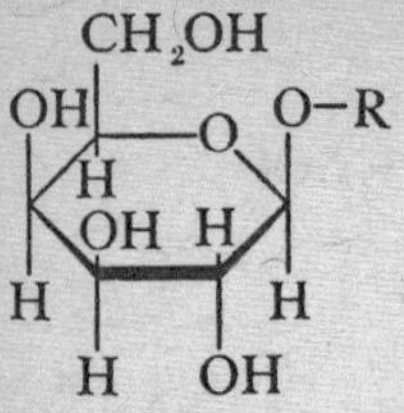

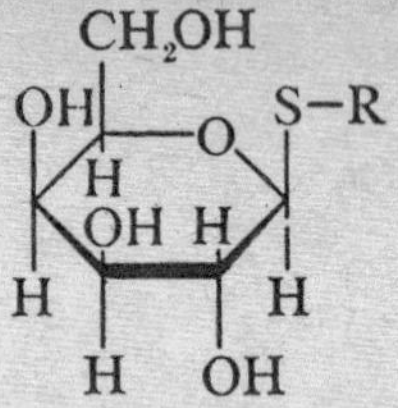

CH$_2$OH   CH$_2$OH

(A)   (B)

It should be remembered that there are many isomers of such compounds (sixteen geometrical isomers, differing by the relative orientation of the OH and H groupings upon carbons 1 to 5; plus the optical opposites of each of these sixteen).

The enzyme in point of fact exactly discriminates between all these isomers, and hydrolyses only one of them. Nevertheless one may 'trick' the enzyme by synthesizing 'steric analogues' of compounds belonging to this series, in which the oxygen of the hydrolysable bond is replaced by sulphur – formula (B) in the above diagram. The sulphur atom, larger than the oxygen, is of the same valency, and for both atoms the orientation of the valencies is the same. The three-dimensional *shape* of these sulphur derivatives is therefore practically the same as that of their oxygen counterparts. But the bond formed by sulphur is much more stable than the oxygen bond. The enzyme consequently fails to hydrolyse these compounds. However, it may be *directly* demonstrated that they form a stereospecific complex with the protein.

Such observations not only confirm the theory of the complex but show that an enzymic reaction is to be considered as made up of two distinct stages:

1. The formation of a stereospecific complex between protein and substrate.

2. The catalytic activation of a reaction within the complex: a reaction *oriented* and *specified* by the structure of the complex itself.

58

This distinction is of major importance, and will enable us to reach one of the central concepts of molecular biology. But first we must note that among the different types of bonds

covalent and
noncovalent bonds

which contribute to the stability of a chemical edifice, two classes can be distinguished: the covalent bond and the noncovalent. Covalent bonds – often referred to as 'chemical bonds' *sensu stricto* – are due to the sharing of electronic orbitals between two or several atoms; noncovalent bonds to several other types of inter- action not implying the sharing of electronic orbitals.

It is not necessary, for our main purposes here, to dwell cn the nature of the physical forces that take part in these different types of interaction. We may start by emphasizing that the two classes of bonds differ from each other through the energy of the associations they ensure. Simplifying some- what, and specifying that we are now considering only those reactions occurring in aqueous phase, we may say that the average amount of energy absorbed or liberated by a reaction involving covalent bonds is in the order of 5 to 20 Kcal per bond. For a reaction involving noncovalent bonds only, the average amount of energy would be between 1 and 2 Kcal.[4]

This considerable difference partially accounts for the difference in stability between 'covalent' and 'noncovalent' chemical constructs. The essential, however, lies not there but in the difference in the so-called 'activation' energies brought into play in the two types of reaction. This point is of highest importance. To clarify it we should recall that a reaction causing a molecular population to pass from a given stable state into another must be understood to include an inter-

4. Let us remember that a bond's energy is, by definition, the energy that must be furnished *to split it*. But in actual fact most chemical – and notably biochemical – reactions consist in the *exchange* of bonds rather than in their outright rupture. The energy brought into play in a reaction is that which corresponds to an exchange of the type $AY + BX \rightarrow AX + BY$. It is therefore always lower than the splitting energy.

mediate state, of potential energy *superior* to either of the two terminal states. The process is often represented by a plot whose abscissa indicates the forward course of the reaction and its ordinate the potential energy (Fig. 1). The difference in potential energy between the terminal states corresponds to the energy released by the reaction; the difference between the initial state and the intermediate ('activated') state is the activation energy: this is the energy that the molecules must *transitorily* acquire in order to enter into reaction. This energy, acquired in the course of the first phase and released in the second, does not figure in the final thermodynamic accounting. However, upon it depends the *speed* of the reaction, which will be practically zero at ordinary temperature, if the activation energy is high. Hence, in order to provoke such a reaction one must either considerably increase the temperature (thereby assuring sufficient energy to the fraction of molecules) or else employ a catalyst, whose role is to 'stabilize' the activated state, thereby reducing the difference of potential between this state and the initial one.

Now – and this is the crucial point – in general:

a. The activation energy of covalent reactions is high; their speed is therefore very slow or zero at low temperature and in the absence of catalysts; while

b. The activation energy of noncovalent reactions is very low if not zero; they therefore occur *spontaneously* and *very rapidly*, at *low temperature*, and in the *absence of catalysts*.

The result is that structures defined by noncovalent interactions can attain a certain stability only if they entail *multiple* interactions. Furthermore, noncovalent interactions acquire a notable amount of energy only when the atoms lie a very short distance apart, practically 'touching' one another. Consequently two molecules (or areas of molecules) will be able to contract a noncovalent association only if the surfaces of both include *complementary sites* permitting several atoms of the one to enter into contact with several atoms of the other.

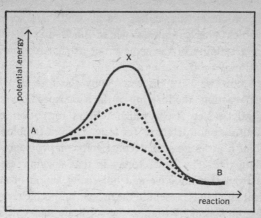

Figure 1. Diagram showing the variation of potential energy of molecules in the course of a reaction. A: initial stable state; B: final stable state; X: intermediate state of potential energy superior to that of the two stable states. Continuous line: covalent reaction; dashes: covalent reaction in the presence of a catalyser which lowers the activation energy; dotted line: noncovalent reaction.

If we now add that the complexes formed between enzyme and substrate are of a noncovalent nature it will be seen why these complexes are *necessarily* stereospecific: they can form only if the enzyme molecule has a site exactly 'complementary' to the shape of the substrate molecule. It will be seen, also, that in the complex the molecule of substrate is necessarily very strictly positioned by virtue of the multiple interactions connecting it to the enzyme molecule's receptor site.

the concept of the noncovalent stereospecific complex

And, lastly, it will be seen that, depending upon the *number* of noncovalent interactions it entails, the stability of a noncovalent complex will vary along a very broad scale. Therein lies a precious property of noncovalent complexes: their stability can be exactly adjusted to the function fulfilled. Enzyme-substrate complexes must be able to assemble and to come apart very rapidly; high catalytic activity demands this.

61

These complexes are indeed easily and very promptly dissociab'e. Other complexes, whose function is permanent, acquire a stability of the same order as that of a covalent association.

Until now we have discussed only the first step in an enzymic reaction: the forming of the stereospecific complex. The catalytic step itself, which follows formation of the complex, need not detain us for long, as from the biological viewpoint it poses no such deeply significant problems as the preceding one. The belief today is that enzymic catalysis results from the inductive and polarizing action of certain chemical groupings present in the protein's 'specific receptor'. Apart from specificity (due to the substrate molecule's very precise positioning vis-à-vis the inducer groups), the catalytic effect is explained by schemes similar to those which account for the action of nonbiological catalysts (such as, notably, H^+ and OH^- ions).

The formation of the stereospecific complex, as a prelude to the catalytic act itself, may therefore be regarded as simultaneously fulfilling two functions:

1. The exclusive *choice* of a substrate, determined by its steric structure.

2. The correct *presentation* of the substrate in the precise position that limits and specifies the catalytic effect of the inducer groups.

The idea of a noncovalent stereospecific complex is applicable not just to enzymes nor even, as will be seen, just to proteins. It is of pivotal importance for the interpretation of all the phenomena of choice, of elective discrimination, that characterize living beings and make them appear to escape the fate pronounced by the second law of thermodynamics. In this connection it is worth glancing again at the example of fumarase.

Using the means of organic chemistry to aminate fumaric acid, one obtains a mixture of the two optical isomers of aspartic acid. The enzyme, on the other hand, catalyses the formation of L-aspartic acid exclusively. This represents an input of information exactly corresponding to a binary choice (since there are two isomers). Here one sees at the most elementary level how structural information can be created and distributed in living beings. The enzyme of course possesses, in the structure of its specific receptor, the information corresponding to this choice. But the energy required to *amplify* this information does not come from the enzyme: to orient the reaction exclusively along one of the two possible paths, the enzyme utilizes the chemical potential constituted by the fumaric acid solution. All the synthesizing activity of cells, however complex, may in the last analysis be interpreted in the same terms.

These phenomena, prodigious in their complexity and their efficiency in carrying out a preset programme, clearly invite the hypothesis that they are guided by the exercise of somehow 'cognitive' functions. The nineteenth-century physicist James Clerk Maxwell attributed such a function to his microscopic demon. We recall how this hypothetical personage, posted at the communicating opening between two enclosed spaces filled with a gas of some kind, was supposed, without any consumption of energy, to manoeuvre an ideal hatch enabling him to prevent certain molecules from passing from one chamber to the other. The gatekeeper could thus 'choose' to allow only fast (high energy) molecules through in one direction, and only slow (low energy) molecules in the other. The result being that, of the two enclosed spaces originally at the same temperature, one grew hotter while the other grew cooler – all without any apparent consumption of energy. However imaginary this experiment, it caused constant confusion among physicists: for it did indeed seem that *through the exercise of his cognitive*

Maxwell's demon

function the demon was able to violate the second law. And as this cognitive function appeared neither measurable nor even definable from the physical standpoint, Maxwell's 'paradox' seemed to defy all analysis in operational terms.

The key to the riddle was provided by Léon Brillouin, drawing upon earlier work by Szilard: he demonstrated that the exercise of his cognitive function by the demon had *necessarily* to entail the consumption of a certain amount of energy which, on balance, exactly offset the lessening entropy within the system as a whole. For the demon to work the hatch 'intelligently', it must first have *measured* the speed of each particle of gas. Now any reckoning – that is to say, any acquisition of information – presupposes an interaction, in itself energy-consuming.

This famous theorem is one of the sources of modern thinking on the equivalence between information and negative entropy. The theorem interests us here precisely because enzymes, at the microscopic level, exercise an order-creating function. But this creation of order, as we have seen, is not gratuitous; it comes about at the expense of a consumption of chemical potential. In short, the enzymes function exactly in the manner of Maxwell's demon corrected by Szilard and Brillouin, draining chemical potential into the processes chosen by the programme of which they are the executors.

We should keep in mind the essential idea developed in this chapter: it is by virtue of their capacity to form, with other molecules, *stereospecific and noncovalent complexes* that proteins exercise their 'demoniacal' functions. The following chapters will illustrate the crucial importance of this key concept, which will recur as the ultimate interpretation of the most distinctive properties of living beings.

4

Microscopic Cybernetics

By virtue of its extreme specificity, an 'ordinary' enzyme (like those taken as examples in the previous chapter) constitutes a completely independent functional unit. The 'cognitive' function of the 'demons' is restricted to the recognition of their specific substrate, to the exclusion both of all other compounds and of anything that may occur within the cell's chemical machinery.

From a glance at a drawing condensing what is now known of cellular metabolism we can tell that even if at each step each enzyme carried out its job perfectly, the sum of their activities could only be chaos were they not somehow interlocked so as to form a coherent system. We have indeed the plainest evidence of the extreme efficiency of the chemical machinery of living beings, from the 'simplest' to the most complex.

functional coherence of cellular machinery

We have of course long been aware of the existence in animals of systems providing large-scale coordination of the organism's performances: that is, the functions of the nervous and endocrine systems. These systems ensure coordination between organs and tissues; that is to say, finally, *among cells*. And we now know that within each cell a cybernetic network almost as complex, if not more so, guarantees the functional coherence of the intracellular chemical machinery – this is what has emerged from studies of only the last twenty years, and even the past five or ten.

We are still far from having analysed in its entirety the system that governs the metabolism, growth, and division of bacteria, the simplest known cells. But thanks to thorough analysis of certain parts of this system, its operative principles are today fairly well understood. It is these principles we shall be discussing in this chapter. We shall see that the elementary control operations are handled by specialized proteins acting as detectors and transducers of chemical information.

regulatory proteins and the logic of regulations

At present the best known of these regulatory proteins are the so-called 'allosteric' enzymes. They compose a special class, by reason of features which distinguish them from 'ordinary' enzymes. Like the latter, allosteric enzymes recognize and bind electively a particular substrate and activate its conversion into products. But these enzymes have the further property of recognizing electively one or several *other* compounds, whose (stereospecific) association with the protein has a modifying effect – that is, depending upon the case, of *heightening or inhibiting its activity with respect to the substrate*.

The regulatory, coordinating function of interactions of this type (known as allosteric interactions) is now proved by countless examples. These interactions may be classified into a certain number of 'regulatory patterns', depending upon the relationship existing between the reaction in question and the metabolic origin of the 'allosteric effectors' controlling it. The main regulatory patterns are these (Fig. 2):

1. *Feedback inhibition*. The enzyme which catalyses the first reaction of a sequence whose end-product is an essential metabolite (a constituent of proteins or of nucleic acids, for

example[1]) is inhibited by the final product of the sequence. The intracellular concentration of this metabolite therefore governs its own rate of synthesis.

2. *Feedback activation.* The enzyme is activated by a product of degradation of the terminal metabolite. This case is frequent with metabolites whose high chemical potential constitutes a source of energy for the cellular machinery. This regulatory pattern thus contributes to maintaining the available chemical potential at a prescribed level.

3. *Parallel activation.* The first enzyme of a metabolic sequence leading to an essential metabolite is activated by a metabolite synthesized by an independent and parallel sequence. This mode of regulation contributes to maintaining a balance between metabolites belonging to the same family and destined for assembly in one of the classes of macro-molecules.

4. *Activation through a precursor.* The enzyme is activated by a compound which is a more or less remote precursor of its immediate substrate. This mode of regulation amounts to keeping the 'demand' subordinate to the 'offer'. One particular but extremely frequent case of this kind is

5. *Activation* of the enzyme *by the substrate* itself. This then plays its own 'ordinary' role and at the same time that of an allosteric effector with respect to the enzyme.

Rarely is an allosteric enzyme subject to only one mode of regulation. As a general rule these enzymes are under the simultaneous control of several allosteric effectors, antago-nistic or cooperative. A frequently encountered situation is a 'ternary' regulation comprising:

a. Activation by the substrate (pattern 5);

b. Inhibition by the end-product of a sequence (pattern 1); and

---

1. Any compound produced by metabolism is called a metabolite; 'essential metabolites' are the compounds universally required for the growth and multiplication of cells.

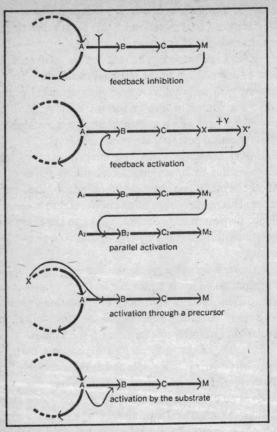

Figure 2. Various 'regulatory modes' assured by allosteric interactions. Arrows with thick lines symbolize reactions producing intermediate compounds (denoted A, B, etc.). The letter M represents the terminal metabolite, conclusion of the sequence of reactions. Fine lines indicate the origin and point of application of a metabolite acting as an allosteric effector, the inhibitor or activator of a reaction.

    c. Parallel activation by a metabolite of the same family as the end-product (pattern 3).

    Here, then, the enzyme simultaneously recognizes all three

effectors and 'measures' their relative concentrations; its activity at any time represents a summing up of these three inputs of information.

To illustrate the refined intricacy of these systems we may mention by way of example the regulatory patterns of 'branching' metabolic pathways (Fig. 3), which are numerous. In these cases, in general, not only are the initial reactions, at the metabolic fork, regulated by retroactive inhibition, but an earlier reaction, higher up on the common branch, is co-governed by the two (or several) final metabolites.[2] The danger of blocking the synthesis of one of the metabolites by an excess of the other is skirted, depending upon the particular case, in one of two different ways, either

a. By delegating to this one reaction two distinct enzymes, each inhibited by one of the metabolites to the exclusion of the other; or

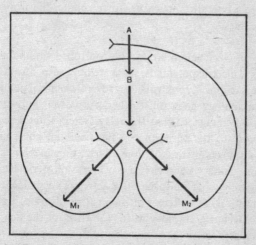

Figure 3. Allosteric regulation of branching metabolic pathways. For the meaning of symbols (letters and arrows) see Figure 2.

2. E. R. Stadtman, *Advances in Enzymology*, 28 (1966), pp. 41–159. G. N. Cohen, *Current Topics in Cellular Regulation*, 1 (1969), pp. 183–231.

b. With a single enzyme, which is only inhibited by the two metabolites acting 'in concert' but not by either one of them alone.

It must be emphasized that, leaving aside the substrate, the effectors which regulate an allosteric enzyme's activity take no part in the reaction itself. With the enzyme they usually form a noncovalent complex, entirely and instantaneously reversible, from which they come away completely unmodified. The consumption of energy incidental to the regulatory interaction is practically nil: it represents but a tiny fraction of the effectors' intracellular chemical potentials. On the other hand the catalytic reaction governed by these very weak interactions may, for its part, involve relatively considerable energy transfers. These systems are thus comparable to those employed in electronic automation circuitry, where the very slight energy consumed by a relay can trigger a large-scale operation, such as, for example, the firing of a ballistic missile.

Just as an electronic relay can be controlled simultaneously by several electric potentials, so, as we have seen, an allosteric enzyme is ordinarily controlled by several chemical potentials. But the analogy goes still further. As is well known, it is usually advantageous that the relay system should respond *nonlinearly* to the variations in the potential governing it; threshold effects are thus obtained, permitting finer regulation. The same holds true in the case of most allosteric enzymes. The curve showing the variation of activity as dependent upon concentration of an effector (including the substrate) is almost always *S*-shaped. In other words the effect of the ligand[3] at first increases *faster* than its concentration. This behaviour is the more remarkable in that it appears to be characteristic of allosteric enzymes. In ordinary 'classic' enzymes, on the

3. We give the name of 'ligand' to a compound defined by its ability to bind to another specific compound.

contrary, the effect always increases *more slowly* than the concentration.

I am not sure what the minimal weight might be for an electronic relay presenting the same logical features as an average allosteric enzyme (receiving and integrating inputs from three or four sources, and responding with threshold effects). Let us say something like a hundredth of a gram. The weight of an allosteric enzyme molecule capable of the same performances is of the order of 10^{-17} of a gram; which is a thousand billion times less than an electronic relay. That astronomical figure affords some idea of the 'cybernetic' (i.e., teleonomic) power at the disposal of a cell equipped with hundreds or thousands of these microscopic entities, all far cleverer than the Maxwell-Szilard-Brillouin demon.

The question is to understand how an allosteric protein performs these extraordinary feats. It is known now that allosteric interactions are mediated by discrete shifts in the protein's molecular shape. In the next

mechanism of allosteric interactions

chapter we shall see that a globular protein's involuted and compact form is stabilized by a host of *noncovalent* bonds which together cooperate in maintaining its structure. This allows certain proteins to assume two (or more) conformational states (just as certain compounds may exist in different allotropic states). The two states in question, and the 'allosteric transition' wherein the molecule shifts from one to the other or back again, are often symbolized thus:

(R)   (T)

Since the shape-recognizing properties of a protein depend upon the shape of its binding site or sites, it may be asserted (and in certain favourable instances directly demonstrated) that these stereospecific properties are modified by the transition. For example: in state R the protein will be able to recognize and therefore to bind ligand α at one site (but not ligand β), whereas in state T it will recognize and bind ligand β but not α. It follows that either ligand will have the effect of stabilizing the protein in this or that of its two states, R or T, at the expense of the other; and that α and β will be mutually antagonistic, since their respective interactions with the protein are mutually exclusive. Now imagine a third ligand γ (which could be the substrate) binding exclusively with the R state in some other site of the molecule than the one where α binds: α and γ, it will be seen, cooperate in stabilizing the protein in its active state (the state which recognizes the substrate). Ligand α and substrate γ will therefore function as activators, ligand β as an inhibitor. The activity of a population of molecules will be proportional to the fraction of them that are in state R, a fraction which will be larger or smaller depending upon the relative concentration of the three ligands as well as upon the value of the intrinsic equilibrium between R and T. Thus the catalytic reaction will be controlled by the magnitude of these three chemical potentials.

Here let us emphasize what is much the most important implication of the above: namely, that the cooperative or antagonistic interactions of the three ligands are *totally indirect. There are no actual interactions between the ligands themselves; all the interactions occur exclusively between the protein and each ligand separately*. We shall return later to this idea, the apparently indispensable key to an understanding of the origin and development of cybernetic systems in living beings.[4]

This scheme of indirect interactions enables us also to

4. J. Monod, J.-P. Changeux, and F. Jacob, *Journal of Molecular Biology*, 6 (1963), pp. 306–29.

account for the subtle refinement shown in the protein's 'nonlinear' response to variations in the concentration of its effectors. All known allosteric proteins are 'oligomeric', made up by the noncovalent assembly of a few (often two or four; less often six, eight, or twelve) chemically identical subunits or 'protomers'. Each protomer bears a receptor for each of the ligands the protein recognizes. As a consequence of its assembly with one or several other protomers the steric structure of each of them is partially 'constrained' by its neighbours. But theory, confirmed by crystallographic evidence, tells us that oligomeric proteins tend to 'pack' in such a way – to adopt such structures – that all the protomers are geometrically equivalent; the constraints they are subjected to are therefore symmetrically distributed between protomers.

Let us now take the simplest case, that of a dimer, and consider what ensues from its dissociation into two monomers. Asunder, they are free to assume a 'relaxed' state, structurally different from the constrained one into which each had been forced when linked together.

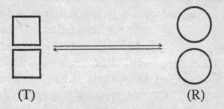

(T)   (R)

The two protomers' change of state, we shall say, is 'concerted'. It is this acting in concert that explains the nonlinearity of response: a ligand molecule's stabilization of the dissociated state R in one of the monomers prevents the other from returning to the associated state, and the same applies for a shift in the opposite direction. The equilibrium between the two states will be a quadratic function of the concentration of the ligands: it would be a fourth-power function for a tetramer, and so on.[5]

I have deliberately confined myself to discussing the simplest possible model, the operations described being present in certain systems which we have reason to regard as 'primitive'. In real systems the dissociation is only rarely complete: the protomers remain associated in both states, though more loosely in one of them. On this basic theme there are many possible variations, but the essential point has been to demonstrate that molecular mechanisms, extremely simple in themselves, account for the 'integrative' properties of allosteric proteins.

Each of the allosteric enzymes referred to up till now constitutes a unit fulfilling a chemical function and at the same time a mediating element in regulatory interactions. Their properties give us an insight into how the homeostatic state of *cellular metabolism* is maintained at a peak of efficiency and coherence.

But by 'metabolism' we essentially mean the transformations of small molecules and the mobilization of chemical potential. In cellular chemistry syntheses proceed upon yet another level: that of the macromolecules, nucleic acids and proteins (including, notably, the enzymes themselves). It has long been known that regulatory systems function at this level too. Their study is much more difficult than that of allosteric enzymes, and as a matter of fact only one of them, thus far, has been thoroughly analysed. We shall take it as an example.

regulation of the synthesis of enzymes

This system, called the 'lactose system', governs the synthesis of three proteins in the bacterium *Escherichia coli*. One

5. J. Monod, J. Wyman, and J.-P. Changeux, *Journal of Molecular Biology*, *12* (1965), pp. 88–118.

of these proteins, galactoside permease, enables the galactosides[6] to penetrate and accumulate within the cell – whose membrane, in the absence of this protein, is impermeable to these sugars. A second protein hydrolyses the β-galactosides. The function of the third protein is not altogether clear and is probably minor. The first and second, on the other hand, are both simultaneously indispensable in the metabolism of lactose (and other galactosides) by the bacteria.

When *Escherichia coli* bacteria grow in a medium devoid of galactosides the three proteins are synthesized at an exceedingly slow rate: about one molecule every five generations. Almost immediately (within about two minutes) after a galactoside – in this connection termed an 'inducer' – is added to the medium, the rate of synthesis of all three proteins increases a thousandfold and maintains this pace as long as the inducer is present. Let the inducer be withdrawn, and within two or three minutes the rate of synthesis slips back to what it was initially.

The conclusions of a long study of this wonderfully and almost miraculously teleonomic phenomenon are summarized in Fig. 4.[7] Here we need not discuss the right-hand part of the diagram, which represents the operation of messenger-RNA synthesis and its translation into polypeptide sequences. Let us simply observe that since the messenger has a rather brief existence (of only a few minutes), its rate of synthesis determines the three proteins' rates of synthesis. Our chief interest is in the components of the regulatory system. They are: the 'regulator' gene, i

the 'repressor' protein, R

the 'operator' segment of the DNA, o

the DNA 'promoter' segment, p

a molecule of inducer galactoside, βG

6. See Chapter III, p. 58.

7. The Finnish scientist Karstrom, who in the thirties made notable contributions to the study of these phenomena, later gave up research, apparently in order to become a monk.

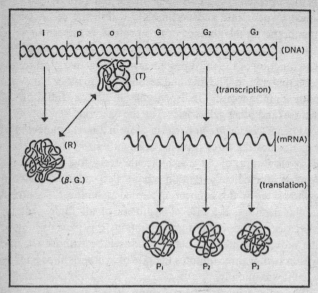

Figure 4. Regulation of the synthesis of the enzymes in the 'lactose system'.

R: repressor-protein, in state of association with the galactoside inducer shown by the hexagon βG.

T: repressor-protein in state of association with operator segment (o) of DNA.

i: 'regulator gene' governing synthesis of the repressor.

p: 'promotor' segment, point of initiation for synthesis of messenger RNA (m RNA).

G_1, G_2, G_3: 'structure' genes governing synthesis of the three proteins in the system, marked P_1, P_2, P_3. (See text, p. 75.)

The regulatory system functions as follows:

1. The regulator gene directs the synthesis, at a constant and very slow rate, of the repressor protein.

2. The repressor specifically recognizes the operator segment to which it binds, forming with it a very stable complex (corresponding to a ΔF of some 15 Kcal).

3. In this state, synthesis of messenger (implying the inter-

vention of the enzyme RNA polymerase) is blocked, presumably by simple steric hindrance, the beginning of this synthesis having to occur at the level of the promoter.

4. The repressor also recognizes β-galactosides, but binds them firmly only when in a free state: hence, in the presence of β-galactosides the operator-repressor complex is dissociated, this permitting the synthesis of messenger and consequently of protein.[8]

It is important to note that both interactions of the repressor are noncovalent and reversible, and that, in particular, the inducer is not modified through its binding to the repressor. Thus the logic of this system is simple in the extreme: the repressor inactivates transcription; it is inactivated in its turn by the inducer. A positive effect, an 'affirmation', results from this double negation. The logic of this negation of the negation, we may add, is not dialectical: it does not result in a new statement but in the reiteration of the original one, written within the structure of DNA in accordance with the genetic code. The logic of biological regulatory systems abides not by Hegelian laws but, like the workings of computers, by the propositional algebra of George Boole.

A great many other similar systems (in bacteria) are known to us today. So far no single one has been entirely dismantled. It appears very likely, though, that the logic of some of them is more complicated than that of the lactose system; in some, for example, negative interactions do not exclusively prevail. But the most general and most significant conclusions to be drawn from the analysis of the lactose system apply as well to these others. It may be said of them all that

a. The repressor, having no activity of its own, is purely a transducer – a mediator – of chemical signals.

8. F. Jacob and J. Monod, *Journal of Molecular Biology*, 3 (1961), pp. 318–56. See also *The Lactose Operon*, Cold Spring Harbor Monograph, ed. J. R. Beckwith and David Zipser (1970).

b. The effect of galactoside upon enzyme synthesis is totally indirect, due exclusively to the repressor's recognition properties and to the fact that two states, each exclusive of the other, are accessible to it. Here again we have what may be termed an allosteric interaction in the general sense discussed earlier.

c. There is no *chemically necessary* relationship between the fact that β-galactosidase hydrolyses β-galactosides, and the fact that its biosynthesis is induced by the same compounds. Physiologically useful or 'rational', this relationship is chemically arbitrary – 'gratuitous', one may say.

This fundamental concept of *gratuity* – i.e., the independence, chemically speaking, between the function itself and the nature of the chemical signals controlling it – applies to

the concept of gratuity

allosteric enzymes. In this case one and the same protein molecule does double duty as specific catalyst and as transducer of chemical signals. But, as we have seen, allosteric interactions are indirect, proceeding exclusively from the protein's discriminatory properties of stereospecific recognition, in the two (or more) states accessible to it. Between the substrate of an allosteric enzyme and the ligands prompting or inhibiting its activity there exists no *chemically necessary* relationship of structure or of reactivity. The specificity of the interactions, in short, has nothing to do with the structure of the ligands; instead, it is entirely due to that of the protein in the various states the protein is able to adopt, a structure in its turn freely and arbitrarily *dictated* by the structure of a gene.

The result – and this is the essential point – is that so far as regulation through allosteric interaction is concerned, *everything is possible*. An allosteric protein should be seen as a specialized product of molecular 'engineering', enabling an interaction, positive or negative, to take place between compounds without chemical affinity, and thereby eventually subordinating any reaction to the intervention of compounds that are chemically foreign and indifferent to this reaction. The way in which allosteric interactions work hence permits a complete freedom in the 'choice' of controls. And these controls, subject to no chemical requirements, will be the more responsive to physiological requirements, by virtue of which they will be selected according to the increased coherence and efficiency they confer on the cell or organism. In short, the very *gratuitousness* of these systems, giving molecular evolution a practically limitless field for exploration and experiment, enabled it to elaborate the huge network of cybernetic interconnections which makes each organism an autonomous functional unit, whose performances appear to transcend, if not to escape, the laws of chemistry.[9]

Actually, as we have seen, when analysed at the microscopic – the molecular – level, these performances appear wholly interpretable in terms of specific chemical interactions, electively assured, freely chosen and organized by regulatory proteins. And it is in the structure of these molecules that one must see the ultimate source of the autonomy, or more precisely, the self-determination that characterizes living beings in their behaviour.

The systems we have studied up to this point are among those that coordinate the cell's activity and make a functional unit of it. In multicellular organisms the coordination between cells, tissues, or organs is guaranteed by specialized systems: not only the nervous and endocrine systems, but also direct interactions between cells. I shall not here discuss the functioning of these systems, which have as yet almost completely evaded microscopic description. We will, however, accept the hypothesis that in these systems the molecular interactions which ensure the transmission and interpretation of chemical signals rest upon proteins endowed with dis-

9. J. Monod, J.-P. Changeux, and F. Jacob, *Journal of Molecular Biology*, 6 (1963), pp. 306–29.

criminatory stereospecific recognition properties, and to which the essential principle of chemical gratuity applies, as it does in the case of allosteric interactions.

To end this chapter, a few words might perhaps be said about the old quarrel between 'reductionists' and 'holists'. Certain schools of thought (all more or less consciously or confusedly

'holism'
v. 'reductionism'

influenced by Hegel) challenge the value of the *analytical* approach to systems as complex as living beings. According to these holist schools which, phoenixlike, are reborn in every generation,[10] the analytic attitude ('reductionist') is doomed to fail in its attempts to reduce the properties of a very complex organization to the 'sum' of the properties of its parts. It is a very stupid and misguided quarrel, which merely testifies to the 'holists' ' total lack of understanding of scientific method and of the crucial role analysis plays in it. How far could a Martian engineer get if, trying to understand an earthly computer, he refused, on principle, to dissect the machine's basic electronic components which execute the operations of propositional algebra? If there is one branch of molecular biology which illustrates better than others the sterility of holist theses as against the cogency of analytical method, it is indeed the study of these microscopic cybernetic systems, which we have briefly considered in this chapter.

The analysis of allosteric interactions reveals, first of all, that teleonomic performances are not the exclusive endowment of complex, multicomponent systems, since *a* protein molecule shows itself capable, not only of electively activating a reaction, but of regulating its activity in response to information emanating from *several* chemical sources.

Secondly, thanks to the concept of gratuity, we see how and

10. Cf. Koestler and Smythies, ed., *Beyond Reductionism* (London, Hutchinson, 1969).

why these molecular regulatory interactions, foiling chemical constraints, succeed in being selectively chosen solely for their contribution to the coherence of the system.

Lastly, the study of these microscopic systems shows us that for complexity, for richness, for potency, the cybernetic network in living beings far surpasses anything that might be predicted by the study of the overall behaviour of whole organisms. And even though these analyses are still far from furnishing us with a complete description of the cybernetic system of the simplest cell, they tell us that, without exception, all activities that contribute to the growth and multiplication of that cell are interconnected and intercontrolled, directly or otherwise.

On such a basis, but not on that of a vague 'general theory of systems',[11] it becomes possible for us to grasp how in a very real sense the organism effectively transcends physical laws – even while obeying them – thus achieving at once the pursuit and fulfilment of its own purpose.

11. Von Bertalanffy, in Koestler and Smythies, op. cit..

5

Molecular Ontogenesis

In both their macroscopic structure and their functions, living beings, as we have seen, are closely comparable to machines. On the other hand, they differ radically from them in the manner of their formation. Any machine or artifact owes its macroscopic structure to the action of external forces, of tools which impose shape upon matter. It is the sculptor's chisel that forms Aphrodite from the block of marble; the goddess herself was born of sea-foam (impregnated by the blood from Uranus' mutilated genitals), whence her body developed of and by itself.

In this chapter I wish to show that this process of spontaneous and autonomous morphogenesis is based on the stereospecific recognition properties of proteins; that it is primarily a microscopic process before manifesting itself in macroscopic structures. Finally, it is in the primary structure of proteins that we shall search for the 'secret' of those cognitive properties thanks to which, like Maxwell's demons, they animate and build living systems.

It should be said at the outset that the problems we are about to tackle, those of the mechanisms of development, still present profound enigmas to biologists. For while embryologists have provided admirable descriptions of development, we are still a long way from knowing how to analyse the ontogenesis of macroscopic structures in terms of microscopic interactions. On the other hand, the construction of certain molecular edifices is now fairly well understood, and the construction process, as I shall try to show, is truly one of 'molecular ontogenesis' in which the physical essence of the phenomenon becomes apparent.

As I indicated earlier, globular protein molecules often appear in the form of aggregates containing a definite number of chemically identical subunits. As this number is usually small, these proteins have been named 'oligomers'. In oligomers the subunits (protomers) are associated by non-covalent bonds. Moreover, as we have already seen, the arrangement of the protomers within the oligomeric molecule is such that each of them is geometrically equivalent to the others. Each protomer, consequently, may be converted into any one of the others by an operation of symmetry – actually by a rotation. It is easily demonstrated that the oligomers so constituted possess the elements of symmetry of one of the rotational point-groups.

Thus these molecules constitute real microscopic crystals. They belong, however, to a special class which I shall call 'closed crystals' for, contrary to ordinary crystals (whose geometry conforms to one of the so-called space-groups), they cannot grow without acquiring new elements of symmetry, while usually shedding some of those they had before.

the spontaneous association of subunits in oligomeric proteins

In Chapter 4 we saw how certain of the functional properties of these proteins are connected with their oligomeric state, including their symmetrical structure. The problem of how these microscopic edifices are constructed is thus quite as significant for biology as it is interesting to physics.

Since the protomers in an oligomeric molecule generally are associated by noncovalent bonds only, it is often possible to separate them into free monomeric units by relatively mild treatment (involving no recourse, for example, to high temperatures or aggressive chemical agents). In this state the protein has generally lost all its functional properties, catalytic or regulatory. However – and this is the important point – if the initial 'normal' conditions are restored (by eliminating the dissociating agent), the subunits will generally reassemble spontaneously, restoring the original 'native' state of the aggregate: the same number of protomers in the same geometrical arrangement, accompanied by the same functional properties as before.

What is more, the reassembly of subunits belonging to a given species of protein will occur not only in a solution containing that one particular protein, but also and just as well in complex 'soups' made up of hundreds, if not thousands, of other proteins. Which is further proof of the existence of an extremely specific recognition process, obviously due to the formation of noncovalent steric complexes interassociating the protomers. This process may be justly considered *epigenetic*[1] since, out of a solution of monomeric molecules devoid of any symmetry, larger molecules, of a higher degree of order, have appeared, and immediately acquired functional properties previously absent.

What chiefly interests us here is the *spontaneous* character of this molecular process of epigenesis. It is spontaneous in two senses:

1. The chemical potential necessary for the forming of the oligomers does not have to be injected into the system: it must be considered as present in the solution of monomers.

---

1. The appearance of new structures and new properties in the course of embryonic development has often been referred to as an 'epigenetic' process, expressive of the gradual enrichment of the organism as it grows from its bare genetic beginnings represented by the initial egg. The adjective is also often employed in reference to now outmoded theories in which the 'preformationists' (who believed that the egg contained a miniature of the adult animal) were lined up in opposition to the 'epigeneticists' (who believed in an *actual* enrichment of the initial genetic information). The term is employed here, not in relation to any theory, but in reference to all processes of structural and functional development.

2. Spontaneous in the thermodynamic sense, the process is also kinetically spontaneous: no catalyst is required to activate it – this, of course, is because the bonds formed are noncovalent. We have already stressed the importance of the fact that both the formation and the splitting of such bonds involve practically nothing in the way of energy.[2]

Such a phenomenon is strictly comparable to molecular crystallization occurring in a solution of component molecules. There, too, order is constituted spontaneously through the association of molecules belonging to a single chemical species. *the spontaneous* The analogy is even more striking *structuration* when, in both cases, structures ar- *of complex particles* ranged according to simple and repetitive geometric rules are seen to take shape. But it has been recently shown that certain organelles much more complex in structure are also the products of spontaneous assembly. This is the case with particles called ribosomes which are the essential components of the mechanism that translates the genetic code, that is, of the protein-synthesizing machinery. These particles, whose molecular weight attains 10^6, are made up by the assembly of some thirty distinct proteins plus three different types of nucleic acids. Although we do not know exactly how these various constituents are disposed within a ribosome, it is certain that their arrangement is extremely precise and that the functioning of the particle depends upon it. Once again, we may witness, *in vitro*, the dissociated constituents of ribosomes spontaneously re-assembling themselves into particles having the same composition, the same molecular weight, the same functional activity as the original 'native' material.[3]

However, the most spectacular example we know so far of

2. See Chapter 3, p. 59.
3. M. Nomura, 'Ribosomes', *Scientific American*, *221* (October, 1969), p. 28.

the spontaneous construction of complex molecular edifices is without doubt that of certain bacteriophages.[4] The complicated and very precise structure of the T4 bacteriophage corresponds to this particle's function, which is not only to protect the genome (i.e., the DNA) of the virus, but to attach itself to the wall of the host cell in order to inject into it, syringelike, its DNA content. The different parts of this microscopic precision machinery can be obtained separately from different mutants of the virus. Mixed together *in vitro* they assemble themselves *spontaneously* to reconstitute particles identical to normal ones and fully capable of exercising their DNA-injecting function.[5]

All these findings are relatively recent, and in this area of research we may expect important advances leading to the *in vitro* reconstitution of more and more complex organelles, such as mitochondria and membranes. The two or three cases mentioned are, however, sufficient to illustrate the process whereby complex structures possessing functional properties develop from the stereospecific, *spontaneous*, assembling of their protein constituents. Order, structural differentiation, acquisition of functions – all these appear out of a random mixture of molecules individually devoid of any activity, any intrinsic functional capacity other than that of recognizing the partners with which they will build the structure. We can no longer speak of crystallization in connection with ribosomes and bacteriophages, since these particles are of a degree of complexity, that is to say of an order, much higher than that of a crystal; but in the last analysis it is nonetheless true that the chemical interactions involved are basically of the same nature as those that construct a molecular crystal. As in a crystal, the structure of the assembled molecules itself

4. Viruses which attack bacteria.

5. R. S. Edgar and W. B. Wood, 'Morphogenesis of bacteriophage T4 in extracts of mutant infected cells', *Proceedings of the National Academy of Science*, 55 (1966), p. 498.

constitutes the source of 'information' for the construction of the whole. These epigenetic processes therefore consist essentially in this: the overall scheme of a complex multimolecular edifice is contained *in posse* in the structure of its constituent parts, but only comes into actual existence through their assembly.

This analysis plainly reduces the old dispute between preformationists and epigeneticists to a quibble over words. The complete structure was never preformed; but the architectural plan for it was present in its constituents themselves, so enabling it to come into being spontaneously and autonomously, without outside help and without the injection of additional information. The necessary information was present, but unexpressed, in the constituents. The epigenetic building of a structure is not a *creation*; it is a *revelation*.

Though admitting that the extrapolation still needs the support of conclusive experimental evidence, modern biologists are convinced that this concept, directly founded upon study of the formation of microscopic edifices, also explains and must be applied to the epigenesis of macroscopic structures (tissues, organs, limbs, etc.). These indeed present problems on a very different scale, in terms both of dimensions and of complexity. Here the most important constructive interactions occur not between molecular components but between cells. It has been established that isolated cells of a given tissue are able to recognize one another discriminatively and to associate; but it is not yet known what components or structures permit cells to identify each other. Everything suggests that the answer lies in the structural characteristics of cellular membranes. But we do not know whether the recognition is of individual molecular shapes or of multi-

microscopic morphogenesis and macroscopic morphogenesis

molecular surface patterns.[6] Whatever the case may be, and even if it is one of patterns not made up of protein components alone, the structure of patterns such as these would of necessity be determined by the shape-recognition properties of their protein components, and also by those of the enzymes responsible for the biosynthesis of a pattern's other components (polysaccharides or lipids, for example).

Thus it may be that the 'cognitive' properties of cells are not the direct but rather a very indirect expression of the discriminatory faculties of certain proteins. Nevertheless the construction of a tissue or the differentiation of an organ – macroscopic phenomena – must be viewed as integrated results of multiple microscopic interactions due to proteins, and as deriving from the stereospecific recognition properties of those proteins, by way of the *spontaneous* forming of noncovalent complexes.

But it must be recognized that this 'reduction to the microscopic' of morphogenetic phenomena is not yet a working theory of those phenomena. It is more a position of principle which specifies only the terms in which such a theory would have to be formulated if it were to aspire to anything better than simple phenomenological description. This principle defines the objective to be reached but throws little light on the way to reach it. One need only consider the formidable problem of accounting in molecular terms for the elaboration of an apparatus as intricate as the central nervous system, requiring thousands of millions of specific interconnections between cells, sometimes sited at relatively great distances apart in the body.

This problem of long-distance influences and orientations is probably the most difficult and the most important in embryology. In their efforts to explain the phenomena of

6. J.-P. Changeux, in 'Symmetry and function in biological systems at the molecular level', *Nobel Symposium* No. 11, ed. A. Engström and B. Strandberg (New York, Wiley, 1969), pp. 235–56.

regeneration, embryologists have introduced the idea of a 'morphogenetic field' or 'gradient', an idea which at first sight seems to go much further than that of a stereospecific molecular interaction within the narrow confines of a few angstroms. However, the latter idea alone makes exact physical sense, and it is by no means inconceivable that a row of such interactions, one triggering the next, could create or define an organization of, for example, millimetric or centimetric proportions. In modern embryology the thinking is along these lines. It is fairly likely that the idea of purely *static* stereospecific interactions will prove insufficient for the interpretation of the morphogenetic field or gradients. It will need the reinforcement of kinetic hypotheses, similar perhaps to those which enable us to interpret allosteric interactions. But personally I am convinced that in the end only the shape-recognizing and stereospecific binding properties of proteins will provide the key to these phenomena.

In analysing the catalytic or regulatory or epigenetic functions of proteins, one is led to recognize that they all depend above all upon the capacities of these molecules for stereospecific association.

According to the thesis put forward in this and the two preceding chapters, all the teleonomic performances and structures of living beings are, at least in principle, analysable in these terms. If this concept is *primary and globular structures of proteins* adequate – and there is no reason to doubt that it is – what is needed to resolve the paradox of teleonomy is a full explanation of the manner in which stereospecific associative protein structures form and of the mechanisms by which they evolve. For the moment I shall consider their manner of formation, and deal with the question of their evolution in later chapters. I hope to show that detailed analysis of these molecular structures, in which the ultimate 'secret' of teleonomy lies hidden, leads to profoundly significant conclusions.

To begin with it must be remembered that the three-dimensional structure of a globular protein is determined by two types of chemical bonds.[7]

1. The so-called 'primary' structure is constituted by a topologically linear sequence of amino acid residues linked by covalent bonds. Thus by themselves these bonds define a fibrous, exceedingly flexible, structure, able in theory to take on an almost infinite variety of shapes.

2. But the so-called 'native' shape of a globular protein is in addition stabilized by a very great number of noncovalent interactions which bind together the amino acid residues distributed along the topologically linear covalent sequence. As a result, the polypeptide fibre folds in a very complex way into a compact pseudo-globular bundle. These complex foldings are what actually determine the molecule's three-dimensional structure, including the exact shape of the stereo-specific binding sites by which the molecule performs its recognition activity. And so one sees that it is the sum or rather the cooperation of a multitude of noncovalent intra-molecular interactions that stabilizes the functional structure of the protein – which in turn enables it to form, electively, stereospecific complexes (likewise noncovalent) with other molecules.

The question concerning us here is the ontogenesis, the origin and development of this special, unique, conformation to which a protein's cognitive function is tied. For a long time it was thought that because of the very complexity of these structures and of the fact that they are stabilized by noncovalent and individually very labile interactions, a vast number of different shapes would be available to a given

7. See Appendix 1, p. 171.

polypeptide fibre. But a considerable body of findings was to show that a given chemical species (defined by its primary structure) exists in its native state, under normal physiological conditions, only in a single conformation (or at the very most in a small number of discrete states, not very different from each other, as is the case with allosteric proteins). This is a very precisely defined conformation, as is proved by the fact that protein crystals yield fine X-ray diffraction images – which means that the position of the great majority of the thousands of atoms composing a molecule is defined to within a fraction of an angstrom. We may add that this combined uniformity and precision of structure is indispensable to specific binding, a biologically essential property of globular proteins.

The formation mechanism of these structures is today well enough understood. We know that:

formation of globular structures

a. The genetic determinism of protein structures *exclusively specifies the sequence* of the amino acid residues corresponding to a given protein; and

b. The polypeptide fibre thus synthesized folds in upon itself *spontaneously* and *autonomously*, ending up in its pseudo-globular functional shape.

Thus, among the thousands of different ways in which the polypeptide fibre could theoretically fold itself, only one is actually adopted. Here we have manifestly a true epigenetic process at the simplest possible level, that of an isolated macromolecule. Thousands of conformations are available to the unfolded fibre. Moreover, before folding it is devoid of any biological activity, and particularly of any capacity for stereospecific recognition. Once folded, on the other hand, it can only take a certain shape, which consequently corresponds to a much higher degree of order. Its functional activity is connected with this state and no other.

The explanation of this little miracle of molecular epigenesis is relatively simple in principle.

1. In the physiologically normal medium, i.e., in aqueous phase, the folded state of the protein is thermodynamically more stable than the unfolded one. The reason for this gain in stability is most interesting and worth noting. About half of the amino acid residues making up the sequence are 'hydrophobic', that is, they behave like oil in water: they tend to collect, freeing the water molecules immobilized through contact with them. As a result the protein assumes a compact structure, by reciprocal contact immobilizing the residues composing the fibre. Whence, for the protein, a heightening of order (negative entropy) – counterbalanced within the system by an attenuation of order (i.e., an increase of entropy) caused by the admixture of the released water molecules.

2. Among the many different folded shapes accessible to a given polypeptide sequence only one, or a very few, will allow the formation of the most compact possible structure. This structure will therefore be favoured above all others. Simplifying a little, we may say that the 'chosen' structure will be the one corresponding to the expulsion of the maximum number of water molecules. Clearly it is upon the relative position – that is, the sequence – of the amino acid residues in the fibre (beginning with the hydrophobic residues) that the various possibilities of achieving compact structure will depend. The globular shape peculiar to a given protein – the special shape required for its functional activity – will therefore be in fact *dictated* by the sequence of residues in the fibre. The important point is that the quantity of information needed to describe the entire three-dimensional structure of a protein is *far greater* than the amount of information defined by the sequence itself. For example, for a polypeptide 100 residues long the information (H) necessary to define the sequence would come to about 2000 bits ($H = \log_2 20^{100}$), whereas to define its three-dimensional structure this sum of

information would have to be supplemented by a great deal more, the exact amount being difficult to calculate (but 1000 to 2000 bits at least).

There may seem to be a contradiction in saying that the genome 'entirely defines' the function of a protein while this function is linked to a three-dimensional structure whose data content is *richer* than the direct contribution made to the structure by the genome. This contradiction has been seized on by certain critics of modern biological theory, in particular by Elsässer, who sees in the epigenetic development of the (macroscopic) structures of living beings a phenomenon beyond physical explanation, by reason of the 'uncaused enrichment' it appears to indicate.

the false paradox of genetic enrichment

Detailed scrutiny of the mechanisms of molecular epigenesis disposes of this objection. The enrichment of information seen in the formation of three-dimensional protein structures is due to the fact that genetic information (represented by the sequence) is expressed under strictly defined initial conditions (aqueous phase, narrow range of temperatures, ionic composition, etc.). The result is that of all the structures possible only one is actually realized. Initial conditions consequently contribute to the items of information finally enclosed in the globular structure. Without specifying it, they contribute to the realization of a unique shape by eliminating all alternative structures, in this way proposing – or rather, imposing – an unequivocal interpretation of a potentially equivocal message.

The structuring process of a globular protein may thus be seen both as the microscopic image and as the source of the autonomous epigenetic development of the organism itself. A development in which several ascending stages or levels are discernible:

1. Folding of the polypeptide sequences culminating in globular structures provided with stereospecific binding properties.

2. Associative interactions between proteins (or between proteins and certain other constituents) so as to build cellular organelles.

3. Interactions between cells, so as to constitute tissues and organs.

4. Throughout the process, coordination and differentiation of chemical activities via allosteric-type interactions.

At each stage more highly ordered structures and new functions appear. Resulting from spontaneous interactions between products of the preceding stages, they reveal successively, like a multistage firework, the latent potentialities of previous levels. The determining cause of the entire phenomenon, its source, is finally the genetic information represented by the sum of the polypeptide sequences, interpreted (or, more exactly, screened) by the initial conditions.

The *ultima ratio* of all the teleonomic structures and performances of living beings is thus enclosed in the sequences of residues making up polypeptide fibres, 'embryos' of the globular proteins which in biology play the role of Maxwell's demons. In a very real sense it is at this level of chemical organization that the secret of life (if there is one) is to be found. And if we could not only describe these sequences but pronounce the law by which they assemble, the secret could be declared open, the *ultima ratio* discovered.

the ultima ratio *of teleonomic structures*

The first description of a globular protein's complete sequence was given by Sanger in 1952. It was both a revelation and a disappointment. This sequence, which was known to define the structure and hence the elective properties of a functional protein (insulin), showed no regularity, special feature, or restrictive characteristic. Even so the hope remained that, with the gradual accumulation of other such findings, a few general laws of assembly as well as certain functional correlations would finally come to light. Today we know hundreds of sequences corresponding to various proteins extracted from all sorts of organisms. From the work on these sequences, and after systematic comparison aided by modern methods of analysis and computing, we can now deduce the general law: it is that of chance. To be more specific: these structures are 'random' in the sense that, even knowing the exact order of 199 residues in a protein containing 200, it would be impossible to formulate any rule, theoretical or empirical, enabling us to predict the nature of the one residue not yet identified by analysis.

To say that in a polypeptide the amino acid sequence is 'random' is not an admission of ignorance. On the contrary, it is a statement of fact, such as, that the average frequency with which such a residue in the polypeptide chain is followed by such an other is equal to the *product* of the average frequencies of each of the two residues in proteins in general. We can illustrate this in another way. Imagine a pack of two hundred cards, each card marked with the name of an amino acid. In the pack the *average* proportion of each of the twenty amino acids is respected. After shuffling, the cards are turned up one by one: the order in which they now appear defines a sequence *which could not be distinguished from a natural one by any objective criterion.*

But while in this sense every primary protein structure looks like the product of a random choice from among the twenty available residues, we must recognize on the other hand that in another, equally significant, sense a given sequence – *this actual one we are dealing with* – has not been synthesized at random: for the very same order it contains is reproduced, practically without error, in all the molecules of the protein under consideration. Were it not so it would be impossible, indeed, to establish the sequence of a population of molecules by chemical analysis.

And so it must be acknowledged that the 'random' sequence in each protein is in fact reproduced thousands and thousands of times over, in each organism, each cell, with each generation, by a highly accurate mechanism which guarantees the invariance of the structure.

Today we know not only the principle but most of the components of this mechanism and they will be discussed in a later chapter. No detailed knowledge of the mechanism is needed to grasp the deep significance *the interpretation* of the mysterious message made up *of the message* by the sequence of residues in a polypeptide fibre. A message which, by every possible criterion, seems to have been composed completely haphazardly; a message nevertheless charged with a meaning which comes out in the discriminative, functional, directly teleonomic interactions of the globular structure: the three-dimensional translation of the ·linear sequence. Globular protein is already at the molecular level, a veritable machine – a machine in its functional properties, but not, we now see, in its fundamental structure, where only the play of blind combinations can be discerned: random chance caught on the wing, preserved, reproduced by the machinery of invariance and thus converted into order, rule, necessity. A *totally* blind process can by definition lead to anything; it can even lead to vision itself. The origin and lineage of the whole biosphere are reflected in the ontogenesis of a functional protein. And the ultimate source of the project that living beings represent, pursue, and accomplish is revealed in this message – in this neat, exact, but essentially indecipherable text formed by primary structure. Indecipherable, since before expressing the physiologically necessary function which it performs spontaneously, it discloses nothing in its structure other than the pure chance of its origin. But for us, this truly is the profounder meaning of this message which comes to us from the most distant reaches of time.

6

Invariance and Perturbations

Ever since its birth in the Ionian Islands almost three thousand years ago, Western philosophy has been divided between two seemingly opposed attitudes. According to one of them the

Plato and Heraclitus

true and ultimate reality of the universe can reside only in perfectly immutable forms, unvarying by essence. According to the other, the

only real truth resides in flux and evolution. From Plato to Whitehead and from Heraclitus to Hegel and Marx, it is clear that these metaphysical epistemologies have always been closely bound up with their authors' ethical and political biases. These ideological edifices, represented as *a priori*, were actually *a posteriori* constructions designed to justify preconceived ethico-political theories.[1]

For science the only *a priori* is the postulate of objectivity, which spares, or rather forbids, it from taking part in the debate. Science studies evolution, whether that of the universe or of the systems it contains, such as the biosphere, including man. We know that any phenomenon, any event, any cognition implies interactions which by themselves generate modifications in the elements of the system. This is not in any way incompatible with the idea that immutable entities exist within the structure of the universe. On the contrary: the

1. See Karl Popper, *The Open Society and Its Enemies* (London, Routledge & Kegan Paul, 1945).

basic strategy of science in the analysis of phenomena is the discovering of invariants. Every law of physics, as for that matter every mathematical development, specifies some invariant relation; science's fundamental statements are expressed as universal 'conservation principles'. It is easily seen, in any example one may like to choose, that it is in fact impossible to analyse any phenomenon in terms other than those of the invariants that are conserved through it. Perhaps the clearest instance of this is the formulation of the laws of kinetics, which *demanded* the invention of differential equations, that is, a means for defining change in terms of what remains unchanged.

It may be asked, of course, whether all the invariants, conservations, and symmetries that make up the texture of scientific discourse are not fictions substituted for reality in order to obtain a workable image, partially emptied of substance, but accessible to the operations of a logic itself founded upon a purely abstract, perhaps 'conventional' principle of identity – a convention with which, however, human reason seems unable to dispense.

This is a classic problem, and I allude to it here in order to point out that its status has been profoundly changed by the 'quantum revolution'. The principle of identity does not figure as a postulate in classical physics. There it is employed only as a logical device, nothing requiring that it be taken to correspond to a substantial reality. It is an altogether different matter in modern physics, one of whose root assumptions is the *absolute* identity of two atoms found in the same quantum state;[2] whence, too, the absolute, nonperfectible representational value that quantum theory assigns to atomic and molecular symmetries. And so today it seems that the principle of identity can no longer be confined to the status

2. V. Weisskopf, in 'Symmetry and function in biological systems at the macromolecular level', *Nobel Symposium* No. 11, ed. Engström and Strandberg (New York, Wiley, 1969), p. 28.

simply of a guideline for argument: it must be accepted as expressing a substantial reality, at least on the quantic scale.

However this may be, there is and will remain a Platonic element in science which could not be taken away without ruining it. Among the infinite diversity of singular phenomena science can only look for invariants.

There was a 'Platonic' ambition in the systematic search for anatomical invariants to which the great nineteenth-century naturalists, after Cuvier and Goethe, devoted themselves. Modern biologists sometimes do less *anatomical* than justice to the genius of the men *invariants* who, behind the bewildering variety of morphologies and modes of life of living beings, succeeded in identifying, if not a unique 'form', at least a finite number of anatomical archetypes, each of them invariant within the group it characterized. It was of course not difficult to see that seals are mammals closely related to carnivores living on land. It was much harder to discern the same fundamental scheme in the anatomy of tunicates and vertebrates, so as to group them together in the phylum Chordata; and it was still more a feat to perceive the affinities between chordates and echinoderms; yet it is certain, and biochemistry confirms it, that sea urchins are much more closely related to us than the members of certain much more evolved groups of invertebrates such as the cephalopods, for example.

It was from this immense research into basic anatomical types that classical zoology and palaeontology were built – a monument whose structure both evokes and justifies the theory of evolution.

Even so, the diversity of types remained, and it had to be recognized that a great many macroscopic structural patterns, radically unlike one another, coexisted in the biosphere. A blue alga, an infusorian, an octopus, and a human being – what had they in common? With the discovery of the cell and the advent of cellular theory a new unity could be seen under this diversity. But it was some time before advances in biochemistry, mainly during the second quarter of the twentieth century, revealed the profound and strict unity, on the microscopic level, of the whole of the living world. Today we know that from the bacterium to man the chemical machinery is essentially the same, in both its structure and its functioning.

1. In its structure: all living beings, without exception, are made up of the same two principal classes of macromolecular components: proteins and nucleic acids. What is more, in all living beings these macromolecules *the chemical* are constituted by the assembling *invariants* of the same residues, finite in number: twenty amino acids for the proteins and four kinds of nucleotides for the nucleic acids.

2. In its functioning: the same reactions, or rather sequences of reactions, are used in all organisms for the essential chemical operations: the mobilization and storing of chemical potential, the biosynthesis of cellular components.

It is true that many variations on this central theme of metabolism are to be met with, each corresponding to a particular functional adaptation. However, they almost always consist in new uses of universal metabolic sequences, hitherto employed for other functions. For instance, the excretion of nitrogen occurs in different forms in birds and mammals: the former excrete uric acid, the latter urea. Now the pathway for the synthesis of uric acid in birds is only a modification, a minor one moreover, of the sequence of reactions which in all organisms synthesizes the so-called purine nucleotides, universal components of nucleic acids. In mammals the synthesis of urea is obtained thanks to a modification of another universal metabolic pathway: that which concludes with the synthesis of arginine, an amino acid present in all proteins. Any number of examples could be given.

It was biologists of my generation who had the excitement of discovering the virtual identity of cellular chemistry throughout the entire biosphere. By 1950 it was established as a certainty, and each new publication added further confirmation of it. The hopes of the most convinced 'Platonists' were being amply fulfilled.

But this gradual disclosure of the universal 'form' of cellular chemistry seemed, in the meantime, to make the problem of reproductive invariance still more acute and paradoxical. If, chemically, the components are the same and are synthesized by the same processes in all living beings, what is the source of their prodigious morphological and physiological diversity? And, still more puzzling, how does each species, using the same materials and the same chemical transformations as all the others, maintain, unchanged from generation to generation, the structural norm that characterizes it and differentiates it from every other?

We now have the solution to this problem. The universal components – the nucleotides on the one side, the amino acids on the other – are the logical equivalents of an alphabet in which the structure and consequently the specific associative functions of proteins are spelled out. All the diversity of structures and performances the biosphere contains can therefore be written in this alphabet. More, with each succeeding cellular generation it is the *ne varietur* reproduction of the text, written in the form of DNA nucleotide sequences, that guarantees the invariance of the species.

The fundamental biological invariant is DNA. That is why Mendel's definition of the gene as the unvarying bearer of hereditary traits, its chemical identification by Avery (confirmed by Hershey), and the elucidation by Watson and

DNA
as the fundamental
invariant

Crick of the structural basis of its replicative invariance, are without any doubt the most important discoveries ever made in biology. To this must be added the theory of natural selection, whose certainty and full significance were established only by those later discoveries.

The structure of DNA; how that structure accounts for its capacity to dictate an exact copy of the nucleotide sequence which specifies a gene; the chemical machinery that translates the nucleotide sequence of a DNA segment into an amino acid sequence in a protein – all these facts and concepts have been thoroughly and well presented for nonspecialists. No detailed review of them need be given here.[3] The following diagram, which only outlines the essence of the two processes of *replication* and of *translation*, will suffice as basis for the present discussion:

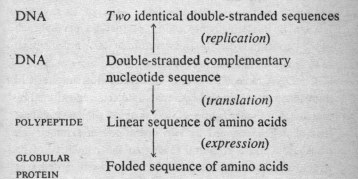

DNA	*Two* identical double-stranded sequences
	↑ *(replication)*
DNA	Double-stranded complementary nucleotide sequence
	↓ *(translation)*
POLYPEPTIDE	Linear sequence of amino acids
	↓ *(expression)*
GLOBULAR PROTEIN	Folded sequence of amino acids

The first point that should be brought out is that the 'secret' of DNA's *ne varietur* replication resides in the *stereochemical*

3. See Appendix 3, p. 181.

complementarity of the *noncovalent* complex constituted by the two strands associated in the molecule. Thus we observe that the fundamental principle of associative stereospecificity, which accounts for the discriminative properties of proteins, is also at the basis of the replicative properties of DNA. But in DNA the complex's topological structure is far simpler than in protein complexes, and it is this which enables the replication mechanism to work. Actually, the stereochemical structure of one of the two strands is entirely defined by the sequence (the succession) of the residues composing it, because *each* of the four residues is *individually* pairable (owing to steric restrictions) with but *one* of the three others. As a result:

1. The steric structure of the complex can be completely represented in *two dimensions*, one of which, finite, contains at each point a pair of mutually complementary nucleotides, while the other contains a potentially infinite sequence of these pairs.

2. Given one – either one – of the two strands, the complementary sequence can be reconstituted step by step by successive additions of nucleotides, each of these being 'chosen' by its sterically predestined partner. So it is that each of the two strands dictates the structure of its complement, so as to reconstitute the entire complex.

The DNA molecule's overall structure is the simplest and most likely to be adopted by a macromolecule constituted by the linear polymerization of identical or similar residues: that of a helix defined by two operations of symmetry, a translation and a rotation. Owing to the regularity of its structure as a whole, the DNA helix may be regarded as a fibrillar crystal. But if one considers the subtler aspects of its structure, it ought rather to be called an *aperiodic* crystal, since the sequence of the base pairs is nonrepetitive. It should be stressed that the sequence is entirely 'free', in the sense that no restriction is imposed upon it by the overall structure, which can accommodate all possible sequences.

As we have just seen, the forming of this structure compares very closely with that of a crystal. Each sequential element in one of the two strands acts the part of a crystalline seed which chooses and orients the molecules that spontaneously link themselves to it, ensuring the crystal's growth. If artificially separated, two complementary strands will *spontaneously* reform the specific complex, each of them almost unerringly choosing its partner from among thousands or millions of other sequences.

However, the growth of each strand implies the formation of *covalent* bonds which sequentially interconnect the nucleotides. The formation of these bonds cannot take place spontaneously: a source of chemical potential and a catalyst are needed. The source of potential is represented by certain bonds present in the nucleotides themselves, which are split in the course of the condensation reaction. The latter is catalysed by an enzyme, DNA polymerase. The sequence, specified by the pre-existing strand, is unaffected by this enzyme. It has been shown, furthermore, that the condensation of mononucleotides activated by nonenzymic catalysers is actually directed by their spontaneous pairing with a pre-existing polynucleotide.[4] Yet it is certain that while the enzyme does not specify the sequence, it does contribute to the precision of the complementary copy – that is, to the fidelity of the transfer of information. As borne out by experiment, it is fidelity of an extremely high degree; but, the process being microscopic, it cannot be absolute. This is a major point, and we shall return to it shortly.

The mechanism whereby the nucleotide sequence is *translated* into an amino acid sequence is a great deal more complicated even in its principle than that of *replication*. Basically, the

4. L. Orgel, *Journal of Molecular Biology*, 38 (1968), pp. 381–93.

latter process is to be explained, as we have just seen, by
direct stereospecific interactions between a polynucleotide
sequence serving as a template and
the translation   the nucleotides that bind thereto. In
of the code   translation noncovalent stereospecific interactions once again guar-
antee the transfer of information. But these governing
interactions contain several successive steps, bringing into
play several components each of which recognizes exclusively
its immediate functional partner. The components involved
at the beginning of this chain of information-transfer enact
their role in complete ignorance of what is 'going on' at the
other end of the chain. Thus, while it is true that the genetic
code is written in a stereochemical language, each of whose
letters consists of a sequence of three nucleotides (a triplet)
in the DNA, specifying one amino acid (among twenty)
in the polypeptide, there exists no direct steric relationship
between the coding triplet and the coded amino acid.

This leads to a most important conclusion: this code,
universal in the biosphere, seems to be chemically *arbitrary*,
inasmuch as the transfer of information could just as well
take place according to some other convention.[5] Indeed,
there are known mutations which, impairing the structure of
certain components of the translation mechanism, thereby
modify the interpretation of certain triplets and thus (with
regard to the convention in force) commit errors which are
exceedingly prejudicial to the organism.

The highly mechanical and even 'technological' aspect of
the translation process merits attention. The successive interactions of the various components intervening at each stage,
leading to the assembly, residue by residue, of a polypeptide
upon the surface of the ribosome, like a milling machine
which notch by notch moves a piece of work through to

5. We shall return to this point in Chapter 8.

completion – all this inevitably recalls an assembly line in a machine factory.

All in all, in the normal organism this microscopic precision machinery confers a remarkable accuracy upon the process of translation. There are mistakes no doubt, but they are so rare that no usable statistics about their normal average frequency are available. The code being unambiguous (for the translation of DNA into proteins), it follows that the sequence of nucleotides in a DNA segment entirely defines the sequence of amino acids in the corresponding polypeptide. Since, as we saw in Chapter 5, the polypeptide sequence specifies completely (under normal initial conditions) the folded structure adopted by the polypeptide once it is constituted, the structural and hence functional 'interpretation' of genetic information is unequivocal, rigorous. No supplementary input of information other than the genetic is necessary; nor, it seems, even possible, as the mechanism as we know it leaves no room for any. And to the extent that all the structures and performances of organisms result from the structures and activities of the proteins composing them, one must regard the total organism as the ultimate epigenetic expression of the genetic message itself.

Another extremely important point should be made: *the translation mechanism is strictly irreversible*. Information is never seen being conveyed in the opposite direction – i.e., from protein to DNA – nor is it *the* conceivable that it could be. This *irreversibility* certitude rests upon an accumulation *of translation* of observations by now so complete and so well verified – and its consequences, especially for evolutionary theory, are so important – that it may be considered one of the fundamental tenets of modern biology.[6] It follows, indeed, that there is no

6. Some critics of the French edition of the present book (Piaget for

possible mechanism whereby the structure and performance of a protein could be modified, and these modifications transmitted even partially to posterity, except by an alteration of the instructions represented by a segment of DNA sequence. Conversely, there is no conceivable mechanism in existence whereby any instruction or piece of information could be transferred to DNA.

Consequently the entire system is totally, intensely conservative, locked into itself, utterly impervious to any 'hints' from the outside world. By its properties, by the microscopic clockwork function that establishes between DNA and protein, as between organism and medium, an entirely one-way relationship, this system obviously defies 'dialectical' description. It is not Hegelian at all, but thoroughly Cartesian: the cell is indeed a *machine*.

It might seem then that by virtue of its very structure this system ought to resist all change, all evolution. This it certainly does, and there we have the explanation[7] for a fact which is indeed far more paradoxical than evolution itself: namely, the prodigious stability of certain species which have been able to reproduce without appreciable modification for hundreds of millions of years.

Physics tells us however that – except at absolute zero, an inaccessible limit – no microscopic entity can fail to undergo quantum perturbations, whose accumulation within a

---

instance) seemed very happy to be able to point to very recent observations as invalidating (so they thought) this statement. This claim rested on the discovery by Temin and by Baltimore of enzymes able to transcribe RNA into DNA, that is, in reverse of the operation of the more usual, already classical, systems. This important observation actually in no way violates the principle that the translation of sequential information from DNA (or from RNA) to protein is irreversible. The authors of the discovery (who are very competent molecular biologists) did not, of course, make any such claim.

7. The partial explanation. See p. 117.

macroscopic system will slowly but surely alter its structure.

Living beings, despite the perfection of the machinery that guarantees the faithfulness of translation, are not exempt from this law. Ageing and death in pluricellular organisms is accounted for, at least in part, by the piling up of accidental errors of translation. These, particularly affecting certain components responsible for the accuracy of translation, tend to precipitate further errors which, ever more frequent, gradually and inexorably undermine the structure of those organisms.[8]

Nor, without violating the laws of physics, could the mechanism of replication be completely immune to disturbances, or accidents. At least some of these disturbances

microscopic
perturbations

create more or less discrete modifications in certain elements of the DNA sequence. Such errors of transcription, thanks to the blind fidelity of the mechanism, will be automatically reproduced. They will be just as faithfully translated into an alteration of the amino acid sequence in the polypeptide corresponding to the DNA segment in which the *mutation* has occurred. But only when this partly new polypeptide has folded in upon itself will the functional import of the mutation become manifest.

In modern biological research some of the work most outstanding in methodology and significance is that known as molecular genetics (Benzer, Yanofsky, Brenner, Crick). This work has, in particular, made it possible to analyse the different types of discrete accidental alterations a DNA sequence may suffer. Various mutations have been identified as due to:

1. The substitution of a single pair of nucleotides for another pair;

8. L. Orgel, *Proceedings of the National Academy of Science*, 49 (1963) p. 517.

2. The deletion or addition of one or several pairs of nucleotides;

3. Various kinds of 'scrambling' of the genetic text by inversion, duplication, displacement, or fusion of more or less extended segments.[9]

We say that these events are accidental, due to chance. And since they constitute the *only* possible source of modifications in the genetic text, itself the *sole* repository of the organism's hereditary structures, it necessarily follows that chance *alone* is at the source of every innovation, of all creation in the biosphere. Pure chance, absolutely free but blind, at the very root of the stupendous edifice of evolution: this central concept of modern biology is no longer one among other possible or even conceivable hypotheses. It is today the *sole* conceivable hypothesis, the only one compatible with observed and tested fact. And nothing warrants the supposition (or the hope) that conceptions about this should, or ever could, be revised.

There is no scientific position, in any of the sciences, more destructive of anthropocentrism than this one, and no other more unacceptable to the intensely teleonomic creatures that we are. So for every vitalist or animist ideology this is the concept or rather the spectre to be exorcized at all costs. It is therefore most important to say something about the words *chance* and *random*, and to specify in what sense they may and must be used with regard to mutations as the source of evolution. The idea of chance is not a simple one, and the word itself is employed in a wide variety of contexts. A few examples will help.

Dice and roulette are called games of chance, and the theory of probability is used to forecast their outcome. But chance enters into these purely mechanical and *macroscopic* games only because of the *practical* impossibility of governing the throw of the dice or the spinning of the little ball with

9. See Appendix 3, p. 181.

sufficient precision. A highly precise mechanical thrower might conceivably be invented which would go far to reduce

operational uncertainty and essential uncertainty

the uncertainty of the outcome. Let us say that in roulette the uncertainty is purely operational and not essential. It is easy to see that the same holds for the theory of numerous phenomena where the concept of chance

and the theory of probability are used for purely methodological reasons.

But in other situations the idea of chance takes on an essential and no longer merely operational meaning. This is the case, for instance, in what may be called 'absolute coincidences', those which result from the intersection of two totally independent chains of events. Suppose that Dr Brown sets out on an emergency call to a new patient. In the meantime Jones the carpenter has started work on repairs to the roof of a nearby building. As Dr Brown walks past the building, Jones inadvertently drops his hammer, whose (deterministic) trajectory happens to intercept that of the physician, who dies of a fractured skull. We say he was a victim of chance. What other term fits such an event, by its very nature unforeseeable? Chance is obviously the essential factor here, inherent in the complete independence of two causal chains of events whose convergence produces the accident.

Now, between the occurrences that can provoke or permit an error in the *replication* of the genetic message and its functional consequences there is also complete independence. The functional effect depends on the structure, on the actual role of the modified protein, on the interactions it ensures, on the reactions it catalyses – all things which have nothing to do with the mutational event itself nor with its immediate or remote causes, regardless of the nature, whether deterministic or not, of those 'causes'.

Finally, on the microscopic level there exists a source of even more radical uncertainty, embedded in the quantum structure of matter. A mutation is in itself a microscopic event, a quantum event, to which the principle of uncertainty consequently applies. An event which is hence and by its very nature *essentially* unpredictable.

The principle of uncertainty was never entirely accepted by some of the greatest modern physicists, including Einstein, who was unwilling to admit that 'God plays at dice'. Certain schools have retained the principle for its operational usefulness but denied it the standing of an essential concept. However, all the efforts made to replace quantum theory by a 'finer' structure from which uncertainty has vanished have ended in failure, and today very few physicists seem disposed to believe that this principle will ever disappear from their discipline.

However this may be it must be stressed that, even were the principle of uncertainty some day abandoned, it would remain true that between the determination, however complete, of a mutation in DNA and the determination of its functional effects on the plane of protein interaction, one could still see nothing but an 'absolute coincidence' like that defined above by the parable of the workman and the physician. The event would still belong to the realm of 'essential' chance. Unless of course we go back to Laplace's world, from which chance is excluded by definition and where Dr Brown was always fated to die knocked out by Jones's hammer.

It will be recalled that Bergson saw evolution as the expression of a creative force, *absolute* in the sense that it was directed to no end except creation in itself and for its own sake. In this he differed radically from the animists (whether Engels, Teilhard de Chardin, or optimistic positivists like Spencer), who all regarded evolution as the majestic unfolding of a programme woven into the very fabric of the universe.

For them, consequently, evolution was not really a creation but uniquely the 'revelation' of nature's hitherto unexpressed designs. Whence the tendency to see in embryonic development an emergence of the same kind as evolutionary emergence. According to modern theory, the idea of 'revelation' applies to epigenetic development, but not of course to evolutionary emergence, which, owing to the fact that it arises from the essentially unforeseeable, is the creator of *absolute* newness. Might this apparent meeting of the ways between Bergsonian metaphysics and scientific thought be yet another effect of sheer coincidence? Perhaps not: artist and poet that he was, and also very well informed on the natural sciences of his day, Bergson could not fail to respond to the dazzling richness of the biosphere and the amazing variety of forms and behaviour it displays, which indeed seem to bear almost direct witness to an inexhaustible, wholly untrammelled creative prodigality.

evolution: absolute creation and not revelation

But where Bergson saw the clearest proof that the 'principle of life' is evolution itself, modern biology recognizes, instead, that all the properties of living beings are based on *a fundamental mechanism of molecular invariance*. For modern theory *evolution is not a property of living beings*, since it stems from the very *imperfections* of the conserving mechanism which indeed constitutes their unique privilege. It must, then, be said that the same source of fortuitous perturbations, of 'noise', which in a nonliving (i.e. nonreplicative) system would gradually lead to the disintegration of all structure, is the progenitor of evolution in the biosphere and accounts for its unrestricted liberty of creation, thanks to the replicative structure of DNA: that registry of chance, that tone-deaf conservatory where the noise is preserved along with the music.

7

Evolution

The initial elementary events which open the way to evolution in the intensely conservative systems called living beings are microscopic, fortuitous, and totally unrelated to whatever may be their effects upon teleonomic functioning.

But once incorporated in the DNA structure, the accident – essentially unpredictable because always singular – will be mechanically and faithfully replicated and translated: that is

chance
and necessity

to say, both multiplied and transposed into millions or thousands of millions of copies. Drawn from the realm of pure chance, the accident enters into that of necessity, of the most implacable certainties. For natural selection operates at the macroscopic level, the level of organisms.

Even today a good many distinguished minds seem unable to accept or even to understand that from a source of noise natural selection could quite unaided have drawn all the music of the biosphere. Indeed natural selection operates *upon* the products of chance and knows no other nourishment; but it operates in a domain of very demanding conditions, from which chance is banned. It is not to chance but to these conditions that evolution owes its generally progressive course, its successive conquests, and the steady development which it seems to suggest.

Some post-Darwinian evolutionists have tended, in discussing natural selection, to propagate a stark, naïvely ferocious idea of it: that of the all-out 'struggle for life' – an expression which, incidentally, is not Darwin's but Herbert Spencer's. The neo-Darwinians of the beginning of this century put forward a much richer concept and showed, on the basis of quantitative theories, that the decisive factor in natural selection is not the struggle for life, but – within a given species – the differential rate of reproduction.

Achievements in contemporary biological research permit a clearer definition of the idea of selection. Of the intra-cellular cybernetic network in particular, of its power, complexity, and coherence, even in the simplest organisms, we have a fairly clear picture; this enables us to understand better than our less well-informed predecessors that any 'novelty', in the shape of an alteration of protein structure, will be tested before all else for its compatibility with the whole of the system already bound by the innumerable controls commanding the execution of the organism's projective purpose. Hence the only acceptable mutations are those which, at the very least, do not lessen the coherence of the teleonomic apparatus, but rather strengthen it in the orientation already assumed or (much more rarely) open up new possibilities for it.

It is the teleonomic apparatus, as it functions when a mutation first expresses itself, that lays down the essential *initial conditions* for the admission, temporary or permanent, or rejection of the chance-bred innovative attempt. It is teleonomic performance, the aggregate expression of the properties of the network of constructive and regulatory interactions, that is judged by selection; and that is why evolution itself seems to be fulfilling a design, seems to be carrying out a 'project', that of perpetuating and amplifying some ancestral 'dream'.

Thanks to the conserving perfection of the replicative apparatus, any mutation, considered individually, is a very rare event. With bacteria – the only organisms for which we have abundant and precise data in this respect – one may say that the probability of a given gene undergoing a mutation which would significantly *the rich resources of chance* affect the functional properties of the corresponding protein is of the order of between one in a million and one in a hundred million per cellular generation. But a population of several thousand million cells can develop in a few millilitres of water. In a population of that size one may be certain that any given mutation will be represented by ten, a hundred, or a thousand samples. One may also estimate the total number of mutants of all kinds in this population at about one hundred thousand to a million.

In so large a population, consequently, mutation is by no means an exceptional phenomenon: it is the rule. And it is within the broader framework of population, not on isolated individuals, that selective pressure is exerted. The population sizes of higher organisms do not, it is true, attain the proportions of bacterial populations; but:

a. In a higher organism, for example in a mammal, the genome contains a thousand times as many genes as the genome of a bacterium; and

b. The number of *cellular* generations, hence the number of chances for mutation, in the germinal line (i.e., the line of cells from ovule to ovule or from spermatozoon to spermatozoon) is very great in a higher organism.

This perhaps accounts for what strikes us as a relatively high incidence of certain mutations in human beings: in the vicinity of 10^{-4} to 10^{-5} for some mutations provoking easily detected genetic infirmities. It should be pointed out that the figures advanced here do not include those individually undetectable mutations which, once associated through sexual recombination, may produce significant effects. Mutations of this sort have probably had a greater importance in evolution than those whose individual effects are more pronounced.

Altogether, we may estimate that in the present-day human population of approximately three thousand million there occur, with each new generation, some hundred thousand million to a billion mutations. This is only to give some idea of the extent of the vast reservoir of fortuitous variability contained within the genome of a species – again in spite of the jealously conservative properties of the replicative mechanism.

Considering the scope of this gigantic lottery and the speed with which nature draws the numbers, it may well seem that the amazing and indeed paradoxical thing, hard to explain,

the 'paradox' of species stability

is not evolution but rather the stability of the 'forms' that make up the biosphere. We know that the anatomical outlines of the main phyla of the animal kingdom were differentiated by the close of the Cambrian period: in other words, five hundred million years ago. It is known, too, that certain species have remained virtually stationary for hundreds of millions of years. The lingula, for example, for the past 450,000,000 years; as for the oyster of 150,000,000 years ago, it looked and probably tasted just like those served in restaurants today.[1] Lastly, one may estimate that the present-day cell, characterized by its invariant basic chemical organization (starting with the structure of the genetic code and the complicated mechanism of translation) has been in existence from two to three thousand million years, provided during all that time with powerful molecular control networks guaranteeing its functional coherence.

The extraordinary stability of certain species, the thousands of millions of years spanned by evolution, the invariance of

1. G. G. Simpson, *The Meaning of Evolution* (New Haven, Yale University Press, 1967).

the cell's basic chemical scheme – these obviously can be explained only by the extreme coherence of the teleonomic system which in evolution has acted as both guide and brake, and has retained, amplified, and integrated only a tiny fraction of the myriad opportunities offered it by nature's roulette.

The replicative system, far from being able to eliminate the microscopic perturbations by which it is inevitably beset, knows only how to register and offer them – almost always in vain – to the teleonomic filter by which their performance is finally judged, through natural selection.

A simple 'point' mutation, such as the substitution of one letter in the DNA code for another, is reversible. Theory predicts this, and experiment proves it. But any appreciable evolution, like the differentiation of two even very nearly related species, is the result of a great many independent mutations successively accumulated in the parent species and then, still at random, recombined thanks to the 'gene flow' promoted by sexuality. Because of the number of independent events that produce it, such a phenomenon is for statistical reasons irreversible.

Evolution in the biosphere is therefore a necessarily irreversible process defining *a direction in time*; a direction which is the *same* as that enjoined by the law of increasing entropy, that is to say, the second law

irreversibility of thermodynamics. This is far more
of evolution than a mere comparison: the second
and the second law law is founded upon considerations
identical to those which establish the

irreversibility of evolution. Indeed, *it is legitimate to view the irreversibility of evolution as an expression of the second law in the biosphere*. The second law, formulating only a statistical prediction, of course does not deny to any macroscopic system the possibility of, almost imperceptibly and for a very brief space, reascending the slope of entropy - taking, as it were, a step backward in time. In living beings it is precisely these fugitive stirrings which, caught up and reproduced by the replicative mechanism, have been retained by selection. In this sense natural selection - based upon a choice of rare and precious incidents contained, among an infinity of others, within the huge reservoir of microscopic chance - constitutes a kind of Wellsian time machine.

It is not surprising but altogether natural that the results obtained by this mechanism for moving backward in time - e.g., the general upward course of evolution, the perfecting and enrichment of the teleonomic apparatus - should appear miraculous to some, paradoxical to others, and that the modern 'Darwinian-molecular' theory of evolution should even today be regarded with suspicion by certain thinkers, philosophers or, for that matter, biologists.

This is due, at least in part, to the extreme difficulty of imagining the inexhaustible resources of the well of chance from which selection draws. Yet a remarkable illustration of it may be found in the organism's *origin of* system of defence through antibodies. *antibodies* These are proteins endowed with the capacity to recognize, by stereospecific association, substances foreign to the organism which have invaded it: for example bacteria and viruses. But as we all know, the antibody that electively recognizes a given substance - for example, a 'steric pattern' peculiar to a certain bacterial species - makes its appearance in the organism (where it will remain present for some time) only after the organism has had at least one experience with the intruder (through vaccination, spontaneous or artificial). It has been further demonstrated that the organism is capable of forming antibodies equipped to cope with practically any natural or synthetic steric pattern. The possibilities, in this respect, seem virtually infinite.

119

For a long time it was thus supposed that the source of information for the synthesis of the antibody's specific associative structure was the antigen itself. Today, however, it is known that the structure of the antibody owes nothing to the antigen. Within the organism specialized cells, produced in great number, possess the property – the unique property – of 'playing roulette' with a well-defined part of the genetic segments that determine the structure of antibodies. The exact functioning of this specialized and ultrarapid genetic roulette has not been entirely elucidated as yet: it seems likely, though, that recombinations as well as mutations occur, but in all cases occur at random, in complete ignorance of the structure of the antigen. The latter then plays the part of selector, differentially favouring the multiplication of those cells which happen to produce an antibody capable of recognizing it.

It is indeed remarkable to find that one of the phenomena occurring in the most exquisitely precise molecular adaptation known to us is based on chance. But it is clear (*a posteriori*) that only such a source as chance could be rich enough to supply the organism with means to repel attack from any quarter.

Another difficulty about accepting the selective theory arises from its having been too often understood or represented as placing the sole responsibility for selection upon conditions of

behaviour orients the pressures of selection

the external environment. This is a completely mistaken conception. For the selective pressures exerted by outside conditions upon organisms are in no case independent of the teleonomic performances characteristic of the species. Different organisms inhabiting the same ecological niche interact in very different and specific ways with external conditions (among which one must include other organisms). These specific interactions, some of which the organism 'elects', determine the nature and orientation of the selective pressure sustained by the organism. Let us say that the 'initial conditions' of selection encountered by a new mutation simultaneously and inseparably include both the environment surrounding it and the total structures and performances of the teleonomic apparatus belonging to it.

It is obvious that the part played by teleonomic performances in the orientation of selection becomes greater and greater, the higher the level of organization and hence *autonomy* of the organism with respect to its environment – to the point where teleonomic performance may indeed be considered decisive in the higher organisms, whose survival and reproduction depend above all upon their behaviour.

It is also evident that the initial choice of this or that kind of behaviour can often have very long-range consequences, not only for the species in which it first appears in rudimentary form, but for all its descendants, even if these constitute an entire evolutionary subgroup. As we all know, the important turning points in evolution have coincided with the invasion of new ecological spaces. If terrestrial vertebrates appeared and were able to initiate the wonderful line from which amphibians, reptiles, birds, and mammals later developed, it was originally because a primitive fish 'chose' to do some exploring on land, where it could however only move about by clumsy hops. This fish thereby created, as a consequence of a change in behaviour, the selective pressure which was to engender the powerful limbs of the quadrupeds. Among the descendants of this daring explorer, this Magellan of evolution, are some that can run at speeds of fifty miles an hour; others climb trees with astonishing agility, while yet others have conquered the air, in a fantastic manner fulfilling, extending, and amplifying the 'dream' of the ancestral fish.

In the evolution of certain groups one observes a general tendency, maintained over millions of years, towards the apparently oriented development of certain organs; this fact shows how the initial choice of a certain kind of behaviour (for example, in the face of attack from a predator) sets the species on the road to a continuous perfecting of the structures and performances which support this behaviour. It is because the ancestors of the horse at an early period chose to live upon open plains and to flee at the approach of an enemy (rather than try to put up a fight or to hide) that the modern species, following a long evolution made up of many stages of reduction, today walks on the tip of a single toe.

We know that certain very precise and complex behaviour patterns, such as the prenuptial ceremonies of birds, are linked to certain particularly conspicuous morphological features. There can be no doubt that this behaviour and the anatomical particularities that go with it evolved *pari passu*, each encouraging and reinforcing the other under the pressure of sexual selection. Once it starts to develop in a species, decorative finery associated with successful mating adds to and confirms the initial pressure of selection and consequently favours any improvement in the finery itself. It is therefore correct to say that sexual drive – or better still, *desire* – created the conditions under which many magnificent plumages were selected.[2]

Lamarck thought that the very strain entailed in an animal's efforts to 'succeed' in life somehow affected its hereditary legacy, entering into it and directly modelling its descendants. The giraffe's immensely long neck would thus express its ancestors' insatiable desire to reach the highest branches of trees. Today this is of course an unacceptable hypothesis; yet one sees that pure selection, operating on elements of behaviour, leads to the result Lamarck wanted to explain: the

2. Cf. N. Tinbergen, *Social Behaviour in Animals* (London, Methuen, 1953).

close interconnection of anatomical adaptations and specific performances.

It is in these terms that the problem of the selective pressures which have oriented human evolution must be considered. It is an exceptionally interesting problem, not just because it is the problem of ourselves, nor because a better insight into the evolutionary roots of our being might give a better understanding of its present nature. An impartial observer, someone from Mars for instance, could not fail to be struck by the fact that the development of symbolic language – man's specific performance and a unique event in the biosphere – opened the way for *another* evolution, creator of a new kingdom: that of culture, of ideas, of knowledge.

This was a unique event: modern linguists dwell on the fact that the symbolic language of human beings is of an utterly different order from the various (auditory, tactile, visual, and other) means of communication animals employ. This is no doubt true. But from there to affirm an absolute break in evolutionary continuity and to claim that human language has owed nothing whatever, even *at the very outset*, to a system of various calls and warnings like those exchanged by apes – this seems to me a difficult step to take, and in any case an unnecessary hypothesis.

language and the evolution of man

Animals, and not only those nearest us on the evolutionary scale, unquestionably possess a brain capable not only of retaining and recording pieces of information but also of associating and transforming them, and of bringing the result of these operations back out in the form of an individual performance; but not – and this is the essential point – in a form which permits the communication to another individual of an original personal association or transformation. But this is what can be done with human language, which may be considered by definition as born on the day when creative combinations – *new* associations achieved by one person – were transmitted to others and no longer had to perish with the individual.

No primitive language exists for us to study: in all the races of our unique modern species, the symbolic instrument has attained roughly the same level of complexity and communicative power. Moreover, according to Chomsky the underlying structure, the 'form' of all human languages, is the same. The extraordinary feats that language both represents and makes possible are obviously connected with the considerable development of the central nervous system in *Homo sapiens*; a development which, for that matter, constitutes his most distinctive anatomical feature.

From what we know of man's most distant ancestors, we can state that his evolution has been marked above all else by the progressive development of the skull, hence of the brain. This has required more than two million years of directed and sustained selective pressure. That pressure must have been heavy, for in evolutionary terms two million years is a relatively short span; and it was *specific*, for nothing similar can be observed in any other line: the cranial capacity of present-day apes is hardly greater than that of their forebears of several million years ago.

Between the privileged evolution of man's central nervous system and that of the unique performance which characterizes him, one must inevitably imagine a closely parallel development, in which language was not only the product but one of the initial conditions of this evolution.

The likeliest hypothesis in my own view is that, appearing very early in our line, the most rudimentary symbolic communication, through its radically new possibilities, was one of those crucial initial 'choices' which decided the future of a species by giving rise to a new selective pressure. This selection must have favoured the development of linguistic ability itself and thus of the organ that served it, the brain. I believe there are powerful arguments for this hypothesis.

The earliest authentic hominids we know of – the australopithecines, whom Leroi-Gourhan rightly prefers to call 'australanthropes' – already possessed, as their defining characteristics, those which separate man from his nearest cousins, the Pongidae (that is to say, the anthropoid apes). The australanthropes had adopted the upright posture, associated not only with specialization of the foot but with many muscular and skeletal modifications, particularly of the vertebral column and the position of the skull in relation to it. Except for the gibbon, every anthropoid moves on all fours; as has often been said, man's evolution must have been tremendously spurred when, standing erect, he freed his hands for other purposes than use in walking. There is no doubt that this invention (a very ancient one, prior to the australanthropes) was of extreme importance: for this alone permitted our ancestors to become hunters able to use their two forelimbs while walking or running.

The cranial capacity of these primitive hominids was, however, scarcely superior to that of a chimpanzee and just below a gorilla's. What the brain can perform is not proportional to its weight; but there is no doubt that its weight imposes limits to intelligence, and that *Homo sapiens* could not have emerged but for the development of his skull.

At any rate it appears established that, although the brain of Zinjanthropus weighed no more than a gorilla's, it was capable of feats unknown among the Pongidae: Zinjanthropus manufactured tools. They were very primitive ones, it is true, and recognizable as artifacts only through the repetition of the same very crude shapes and their location in the vicinity of certain fossil remains. The larger apes used natural 'tools' – stones or branches of trees – when occasion arose, but they produced nothing comparable to artifacts fashioned according to a recognizable *norm*.

Chance and Necessity

Thus Zinjanthropus must be considered a very primitive *Homo faber*. Now it seems likely that there must have been a close correlation between the development of language and that of an industry showing purposive and disciplined activity.[3] Hence it would be reasonable to suppose that the australanthropes possessed an instrument of symbolic communication proportionate to their rudimentary industry. Further, if it is true, as Dart believes,[4] that the australanthropes successfully hunted such powerful and dangerous beasts as the rhinoceros, the hippopotamus, and the panther, they must have hunted as a group carrying out a previously concerted project. Language would have been required for its preliminary formulation.

This hypothesis is not vitiated by the fact that the australanthropes' brain was of very modest size. For recent experiments with a young chimpanzee seem to show that, while apes are incapable of learning spoken language, they can assimilate and use some elements of the sign language employed by deaf-mutes.[5] Hence there are grounds for supposing that the acquisition of the power of articulate symbolization might have followed upon some not necessarily very elaborate neurophysiological modifications in an animal which at this stage was no more intelligent than a present-day chimpanzee.

But it is evident that, once having made its appearance, language, however primitive, could not fail greatly to increase the survival value of intelligence, and so to create a formidable and oriented selective pressure in favour of the development of the brain, pressure which could never be experienced by a dumb species. As soon as a system of symbolic communication

3. Leroi-Gourhan, *Le Geste et la Parole* (Paris, Albin-Michel, 1964); R. L. Holloway, *Current Anthropology*, *10* (1969), p. 395; J. Bronowski, in *To Honor Roman Jakobson* (Paris, Mouton, 1967), p. 374.

4. Cited by Leroi-Gourhan, op. cit.

5. B. T. Gardner and R. A. Gardner, in *Behavior of Non-Human Primates*, ed. Schrier and Stolnitz (New York, Academic Press, 1970).

came into being, the individuals, or rather the groups best able to use it, acquired an advantage over others incomparably greater than any that a similar superiority of intelligence would have conferred on a species without language. We see too that the selective pressure engendered by speech was bound to steer the evolution of the central nervous system in the direction of a special kind of intelligence: the kind most able to exploit this particular, specific performance with its immense possibilities.

This hypothesis would be little more than attractive and reasonable if it were not justified by certain linguistic evidence being compiled today. The study of children's acquisition of

the primary acquisition of language

language irresistibly suggests that this astonishing process is by its very nature profoundly different from the orderly apprenticeship of a system of formal rules.[6] The child learns no rules, and he does not try to imitate adult speech. One might say that he takes from it whatever suits him at each stage of his linguistic development. At the earliest stage (toward eighteen months of age) the child may have a stock of some ten words, which he uses separately, without ever associating them even by imitation. Later he will combine words two, three, and more at a time, according to a syntax which again is not a mere repetition or copying of adult language. This process is, it seems, universal and its chronology the same for all tongues. It is difficult for an adult observer to believe how easily the child masters language after two or three years of this playing with it.

One is led to assume that all this must reflect an embryological, an epigenetic, process in the course of which the neural structures underlying linguistic performances develop. This assumption is borne out by observations on trauma-provoked

6. E. Lenneberg, *Biological Foundations of Language* (New York, Wiley, 1967).

aphasias. The younger the child in whom these aphasias occur, the more quickly and completely they tend to regress. But the impairment becomes irreversible if the lesions occur at the approach of puberty or later. Besides these, a considerable body of findings confirms that there is an age which is critical for the spontaneous acquisition of language. As everybody knows, to learn a second language demands a great deal of determined and systematic effort in an adult; and the status of the language thus learned practically always is, and remains, inferior to that of the native tongue, spontaneously acquired.

Anatomical evidence confirms the idea that the primary acquisition of language is bound up with a process of epigenetic development. It is known that the maturing of the brain continues after birth but halts *the acquisition* at puberty. This development seems *of language* to consist mainly in a considerable *programmed in* amplification of the network of inter- *the epigenetic* connections between cortical neurons. *development* Very rapid during the first two years, *of the brain* the process later slows down. It does not (visibly) extend beyond puberty and therefore coincides with the 'critical period' during which primary acquisition is possible.[7]

From this it is a small step, which I personally am prepared to take, to the theory that if, in the child, the acquisition of language appears so miraculously spontaneous it is because it is an integral part of an epigenetic development *one of whose functions is to prepare for language*. To be a little more precise: the development of the cognitive function itself depends, beyond any doubt, upon this postnatal growth of the cortex. It is the acquisition of language in the course of this epigenesis that makes for its association with the cognitive function – an association so intimate that it is very

7. Ibid.

difficult to separate, by introspection, the utterance from the thought it expounds.

It is generally agreed that language is no more than a 'superstructure', as indeed it would seem in the light of the great diversity of human languages, products of the second evolution, that of culture. However, the extent and refinement of the cognitive functions in *Homo sapiens* clearly have their *raison d'être* in and through language alone. Deprived of this instrument, they are for the most part unusable, paralysed. Thus approached, the capacity for language can no longer be regarded as a superstructure. It must be conceded that, between the cognitive functions and the symbolic language they beget – and through which they are articulated – there is in modern man a close symbiosis which can only be the product of lengthy common evolution.

According to Chomsky and his school, linguistic analysis in depth reveals one basic 'form' common to all human languages, beneath their boundless diversity. So Chomsky says this form must be considered *innate* and characteristic of the species. Certain philosophers or anthropologists have been shocked by this thesis, and see it as a return to Cartesian metaphysics. Provided its implicit biological content be accepted I see nothing whatever wrong with it. On the contrary, it strikes me as a most natural conclusion, once one assumes that the evolution of man's cortical structures could not but be influenced by a capacity for language acquired very early and in the crudest possible state. This amounts to assuming that spoken language, when it appeared among primitive mankind, not only made possible the evolution of culture but contributed decisively to man's *physical* evolution. If these are correct assumptions, the linguistic capacity revealed in the course of the brain's epigenetic development is today part of 'human nature', itself defined within the genome in the radically different language of the genetic code. A miracle? Certainly, since in the final analysis language too was a product of chance. But the day Zinjanthropus or one of his friends first used an articulate symbol to represent a category, he immensely increased the probability that a brain might one day emerge capable of conceiving the Darwinian theory of evolution.

8

The Frontiers

When one thinks about the tremendous journey of evolution over the past three thousand million years or so, the pro-
the present frontiers of knowledge in the field of biology
digious wealth of structures it has engendered, and the extraordinarily effective teleonomic performances of living beings, from bacteria to man, one may well wonder whether all this might not be the product of a vast lottery, in which natural selection has blindly picked the rare winners from among numbers drawn at utter random.

Although a detailed review of the accumulated modern evidence assures us that this conception alone is compatible with the facts (notably with the molecular mechanisms of replication, mutation, and translation), it affords no synthetic, intuitive, and immediate understanding of evolution as a whole. The miracle is 'explained'; it does not strike us as any less miraculous. As Françcis Mauriac wrote, 'What this professor says is far more incredible than what we poor Christians believe.'

This is true, just as it is true that we cannot obtain a satisfactory mental image of certain abstractions in modern physics. But we also know that such difficulties cannot be taken as arguments against a theory which is supported by experiment and logic. In the case of physics, microscopic or cosmological, we can see what the trouble is: the scale of the envisaged phenomena transcends the categories of our immediate experience. Only abstraction can supply this deficiency, but without curing it. In the case of biology the difficulty is of another order. The elementary interactions upon which everything hinges, thanks to their 'mechanical' character, are relatively easy to grasp; it is the phenomenal complexity of living systems which defies intuitive global representation. But in biology, as in physics, these psychological difficulties are not an argument against theory and observation.

Today it may be said that the elementary mechanisms of evolution have been not only understood in principle but exactly identified. The solution found is the more satisfactory since the mechanisms involved are those that ensure the stability of species: replicative invariance in DNA and teleonomic coherence in organisms.

The evolutionary concept is central to biology and will be defined and elaborated for many years to come. But in essence, the problem has been solved, and evolution is now well within the frontiers of knowledge.

The present challenge, as I see it, is in the areas at the two extremes of evolution: the origin of the first living systems, on the one hand; on the other, the inner workings of the most intensely teleonomic system ever to have emerged, to wit, the central nervous system of man. In this chapter I shall try to chart these two borderlands of the unknown.

It might be thought that the discovery of the universal mechanisms basic to the essential properties of living beings would have helped solve the problem of life's origins. As it turns out, these discoveries, by almost entirely transforming the question (today posed in much more precise terms), have shown it to be even more difficult than it formerly appeared.

Three presumptive stages in the process which led to the

emergence of the first organisms may *a priori* be distinguished:

1. The formation on earth of the main chemical building blocks of living beings: nucleotides and amino acids.

the problem of life's origins

2. The formation, from these materials, of the first macromolecules capable of replication.

3. The evolution which elaborated a teleonomic apparatus around these 'replicative structures', eventually leading to the primitive cell.

Different problems arise in the interpretation of each of these stages. The first, often called the 'prebiotic' phase, is open to theoretical and indeed to experimental study. While uncertainty remains, and will doubtless continue, as to the paths actually followed by prebiotic chemical evolution, the overall picture seems fairly clear. Four thousand million years ago conditions in the atmosphere and on the earth's surface favoured the accumulation of certain simple carbon compounds such as methane. There was also water and ammonia. Now from these simple compounds and in the presence of nonbiological catalysts it is fairly easy to obtain numerous more complex compounds, among which figure some amino acids and some precursors of nucleotides (nitrogenous bases, sugars). Remarkably enough, under certain altogether plausible sets of conditions, these syntheses yield a very high percentage of compounds identical or analogous to those which enter the make-up of the modern cell.

And so it may be considered as *proved* that at a given moment in the earth's history certain bodies of water *could* have contained in solution high concentrations of the essential components of the two classes of biological macromolecules, nucleic acids and proteins. In this prebiotic 'soup' various macromolecules might have formed, through polymerization of their precursors, amino acids and nucleotides. In the laboratory, under 'plausible' conditions, some polypeptides

133

and polynucleotides similar in general structure to 'modern' macromolecules might have actually been obtained.

Hence, no major difficulties up to this point. But the decisive step from the first stage to the second has yet to be taken: the formation of macromolecules capable, under the conditions prevailing in the primordial soup, of promoting their own replication unaided by any teleonomic apparatus. This difficulty does not seem insurmountable. It has been demonstrated that a polynucleotide sequence is effectively able to guide, by spontaneous base-pairing, the synthesis of the complementary sequence. Obviously, such a mechanism could only have been very inefficient and subject to innumerable errors. But from the moment it got under way, the three fundamental processes – replication, mutation, and selection – were at work and must have given a considerable advantage to the macromolecules best able, by their sequential structure, to replicate spontaneously.[1]

The third step, according to our hypothesis, was the gradual emergence of teleonomic systems which, around replicative structures, were to construct an *organism*, a primitive cell. It is here that we reach the real 'sound wall', for we have no idea what the structure of a primitive cell might have been. The simplest living system known to us, the bacterial cell, a tiny piece of extremely complex and efficient machinery, attained its present state of perfection perhaps a thousand million years ago. Its overall chemical plan is the same as that of all other living beings. It employs the same genetic code and the same mechanism of translation as do, for example, human cells.

Thus the simplest cells available to us for study have nothing 'primitive' about them. Selection operating over five hundred or a thousand milliard generations has left them with a teleonomic apparatus so powerful that no vestiges of truly primitive structures are discernible. Without the help of fossils, such an evolution cannot possibly be reconstructed.

1. L. Orgel, *Proceedings of the National Academy of Science, 49* (1963).

Still, one would like at least to try to suggest a plausible hypothesis as to the route this evolution followed, and especially as to its starting point.

The development of the metabolic system, which, as the primordial soup thinned, must have 'learned' to mobilize chemical potential and to synthesize the cellular components, poses Herculean problems. So does the emergence of the selectively permeable membrane without which there can be no viable cell. But the major problem is the origin of the genetic code and of its translation mechanism. Indeed, it is not so much a 'problem' as a veritable enigma.

The code is meaningless unless translated. The modern cell's translating machinery consists of at least fifty macromolecular components *which are themselves coded in DNA: the code cannot be translated except by products of translation*. It is the modern expression of *omne vivum ex ovo*. When and how did this circle

the riddle of the code's origins

become closed? It is exceedingly difficult to imagine. But the fact that the code is now deciphered and known to be universal at least allows us to frame the problem in precise terms; simplifying somewhat, in those of the following alternatives. Either:

a. Chemical – or, to be more exact, stereochemical – reasons account for the structure of the code; if a certain codon was 'chosen' to represent a certain amino acid it is because there existed a certain stereochemical affinity between them; or else

b. The code's structure is chemically arbitrary: the code as we know it today is the result of a series of random choices which gradually enriched it.

Hypothesis *a* seems far the more appealing. First, because it would explain the universality of the code. Next, because it permits us to imagine a primitive translation mechanism in which the sequential aligning of amino acids to form a polypeptide would be due to a direct interaction between the amino acids and the replicative structure itself. Finally, and above all, because in principle this hypothesis, if true, would be verifiable. Numerous attempts to verify it have in fact been made: on the whole they have up to now proved negative.[2]

Perhaps we have not yet heard the last word on this score. Pending the unlikely confirmation of this first hypothesis we are reduced to the second, displeasing from the methodological viewpoint – which does not by any means signify that it is incorrect. Displeasing for several reasons: it does not explain the code's universality, and it must be assumed that out of many efforts at elaboration only one survived. This in itself makes sense, but does not provide us with any model of primitive translation. Here speculation must take over, and many very ingenious ideas have been put forward: the field is only too open.

The enigma remains, masking the answer to a question of profound interest. Life appeared on earth: what, *before the event*, were the chances that this would occur? The present structure of the biosphere certainly does not exclude the possibility that the decisive event occurred *only once*. Which would mean that its *a priori* probability was virtually zero.

This idea is distasteful to most scientists. Science can neither say nor do anything about a unique occurrence. It can only consider events which form a class, whose *a priori* probability, however faint, is definite. Now through the very universality of its structures, starting with the code, the biosphere looks like the product of a unique event. It is possible of course that its uniform character was due to elimination through selection of many other attempts or variants. But nothing compels this interpretation.

Among all the events possible in the universe the *a priori* probability of any particular one of them occurring is next to zero. Yet the universe exists; particular events must occur

2. See F. Crick, *Journal of Molecular Biology*, 38 (1968), pp. 367–79.

in it, the probability of which (before the event) was infinitesimal. At the present time we have no justification for either asserting or denying that life made only one single appearance on earth, and that, as a consequence, before it appeared its chances of occurring were almost nil.

This idea is displeasing to biologists, not only because they are scientists. It offends our very human tendency to believe that everything real in the world is necessary, and rooted in the very beginning of things. We must be constantly on guard against this notion, this powerful feeling of destiny. Immanence is alien to modern science. Destiny is written as and while, not before, it happens. Our own was not written before the emergence of the human species, the only one in all the biosphere to use a logical system of symbolic communication – another event, which by its very uniqueness should warn us against any anthropocentrism. If it was unique, as the appearance of life itself may have been, it was because before it did appear its chances of doing so were almost nonexistent. The universe was not pregnant with life nor the biosphere with man. Our number came up in the Monte Carlo game. Is it surprising that, like the person who has just made a million at the casino, we should feel strange and a little unreal?

A logician might remind the biologist that his efforts to 'understand' the complete functioning of the human brain are doomed to fail, since no logical system can produce an integral description of its own structure. *the other* ture. This warning would be some-*frontier:* what inappropriate, considering how *the central* far we still are from that ultimate *nervous system* borderline of knowledge. At any rate this logical objection does not apply to the analysis by man of the central nervous system of an animal, a system which may be supposed to be less complex and less powerful than our own. But even in this case a major difficulty remains: an animal's conscious experience is and no doubt will always be impenetrable to us. As long as this is so, it is questionable whether any exhaustive description of the workings of, say, the brain of a frog is basically possible. Nothing will ever be a suitable substitute for the exploration, however restricted, of the human brain, because this alone enables us to compare objective experimental data with the facts of subjective experience.

In any case, the structure and functioning of the brain can and must be explored simultaneously at every accessible level in the hope that these investigations, very different both in their methods and in their immediate object, will some day converge. At present they only converge in the difficulty of the problems they all raise.

Among the knottiest and most important of these problems are those surrounding the epigenetic development of a structure as complex as the central nervous system. In man it contains from one to ten billion neurons interconnected by means of about a hundred times as many synapses, some of which connect nerve cells lying far apart from each other. I have already mentioned the enigma presented by the establishment of long-distance morphogenetic interactions. Such problems can at least be clearly posed, thanks notably to some remarkable experimental work.[3]

To understand the functioning of the central nervous system we must know that of the synapse, its primary logical element. Investigation is easier here than at any other level, and refined techniques have yielded a considerable mass of findings. However, we are still a long way from an interpretation of synaptic transmission in terms of molecular interaction. Yet that is a most essential question, for therein probably lies the ultimate secret of memory. It was long ago proposed that the memory trace should be registered in the

3. R. W. Sperry, *passim.*

form of a more or less irreversible alteration of the molecular interactions responsible for transmitting the nerve impulse through synapses. This theory is plausible, but has not been directly proved.[4]

Despite this profound ignorance concerning the fundamental mechanisms of the central nervous system, modern electrophysiology, examining the integration of nerve signals, especially in certain sensory pathways, has produced highly significant results.

First of all, with respect to the properties of the neuron as integrator of the signals it may receive, through the intermediary of synapses, from numerous other cells; analysis has shown that in its performances the neuron closely resembles the integrated components of an electronic computer. For example, like the latter, it is capable of carrying out all the logical operations of propositional algebra; it can also add or subtract different signals while taking into account their coincidence in time; it can modify the *frequency* of the signals it transmits in keeping with the *amplitude* of those it receives. In fact, it seems as though no unitary component being used nowadays in modern computers is capable of such varied and finely modulated performances. Between cybernetic machines and the central nervous system the analogy remains impressive and the comparison fruitful; but we must note that, at present, the parallel is confined to the lower levels of integration: the initial stages of sensory analysis, for example. The higher functions of the cortex, which achieve expression through language, seem to elude such methods of approach. One may wonder whether the difference is 'quantitative' (a greater degree of complexity) or 'qualitative'. In my view,

4. A theory according to which the memory is coded in the sequences of residues of certain macromolecules (ribonucleic acids) recently found acceptance among certain physiologists. They apparently believed they had thereby linked up with and could borrow from the concepts derived from the study of the genetic code. However, this theory is untenable in the light, precisely, of what we now know about the code and the mechanisms of translation.

this question has no meaning. Nothing warrants the supposition that the basic interactions are different in nature at different levels of integration. But if there is a case where the first law of dialectics is applicable, this indeed is it.

The very refinement of the cognitive functions in man, and their copious applications, mask the prime functions performed by the brain in the animal series (to which man belongs). These prime functions may

functions of the central nervous system

be listed and defined in the following way:

1. To control and coordinate neuromotor activity, particularly in accord with sensory inputs.

2. To contain, in the form of genetically determined elements of circuitry, more or less complex programmes of action, and to set them in motion in response to particular stimuli.

3. To analyse, sift, and integrate sensory inputs so as to obtain a representation of the outside world geared to the animal's specific performances.

4. To register events which, by the yardstick of those specific performances, are significant; to group them into classes according to their analogies; to associate these classes according to the relationship (of coincidence or succession) of the events constituting them; and to enrich, refine, and diversify the innate programmes by incorporating these experiences in them.

5. To imagine, that is to say, *represent* and *simulate* external events and programmes of action for the animal itself.

The functions noted in the first three paragraphs are fulfilled by the central nervous system of, for example, the Arthropoda, which are not usually reckoned among the higher animals. The most spectacular examples known of very complex innate programmes of action are found in insects. It is doubtful whether among these creatures the functions outlined in paragraph (4) play an important role;[5] on the other hand, they contribute in a most important manner to the behaviour of the higher invertebrates, such as the octopus,[6] and of course to that of all the vertebrates.

As for the functions in paragraph (5), which we may term 'projective', they are probably the prerogative of the higher vertebrates, perhaps of mammals only. But here consciousness becomes a barrier, and it may be that we can perceive the outward signs of this activity (dreaming, for example) only in our nearer cousins, without it being totally absent in other species.

The functions cited under (4) and (5) are cognitive, while those in paragraphs (1), (2), and (3) are solely coordinative and representational. Only the functions in the last paragraph can be creative of *subjective experience*.

According to the proposition in paragraph (3), analysis by the central nervous system of sense impressions furnishes a meagre and slanted image of the external world; a kind of résumé where the emphasis and focus *the analysis* are exclusively upon what is of special *of sense* interest to the animal, in view of its *impressions* specific behaviour. (This is basically a 'critical' résumé, the word being taken in a complementary acceptation of the Kantian sense.) Experiment abundantly bears this out. For instance, the analyser located behind a frog's eye permits it to see a fly (i.e., a black speck) that is moving, but not a fly at rest;[7] and so the frog will only catch its prey in flight. We must stress, and electrophysiological investigation has proved, that this behaviour does not indicate the frog's disdain for a motionless black speck which does not definitely mean food. The

5. Save perhaps in the case of bees.
6. J. Z. Young, *A Model of the Brain* (Oxford, Clarendon Press, 1964).
7. H. B. Barlow, *Journal of Physiology*, *119* (1953), pp. 69–88.

image of the motionless speck is certainly registered upon the frog's retina; but it is not *transmitted*, the system being excited only by a moving object.

Certain experiments upon cats[8] suggest an interpretation of the strange fact that a field simultaneously reflecting all the colours of the spectrum is seen as a *white* expanse, although white is subjectively interpreted as complete absence of colour. These experimenters have shown that, due to cross inhibitions between certain neurons responding to various wavelengths, these neurons do not send signals when the retina is uniformly exposed to the entire gamut of visible wavelengths. So, in a subjective sense, Goethe was right in his argument with Newton. An error wholly forgivable in a poet.

Neither is there any doubt that animals are able to classify objects or relationships between objects according to abstract categories, particularly geometrical ones: an octopus or a rat can learn to distinguish such figures as a triangle, circle, or square, and recognize them unfailingly by their geometrical features, regardless of size, orientation, or the colouring of the real object presented to them.

Study of the circuits which analyse the figures placed in a cat's field of vision shows that these recognitions of geometry are due to the actual structure of the circuits that filter and recompose the retinal image. Actually, these analysers impose a restrictive grid upon the image, from which they extract certain simple elements. Certain nerve cells, for example, respond only to the figure of a straight line sloping down from left to right; others to a line sloped in the opposite direction. Thus it is not so much that a clear geometrical 'idea' is conveyed by the image of the object; rather, the sense analyser perceives and recomposes the object out of its simplest geometrical elements.[9]

8. T. N. Wiesel and D. H. Hubel, *Journal of Neurophysiology, 29* (1966), pp. 1115–56.

9. D. H. Hubel and T. N. Wiesel, *Journal of Physiology, 148* (1959), pp. 574–91.

The Frontiers

These contemporary discoveries therefore, in a new sense and a different context, support Descartes and Kant, against the uncompromising empiricism which has dominated science for the past two hundred years, casting suspicion on any hypothesis positing the innateness of cognitive frames of reference. Certain contemporary ethologists still seem attached to the idea that the elements of animal behaviour are some of them innate, some learned, the one mode of acquisition being strictly separate from and absolutely excluding the other. This conception is completely mistaken as has been vigorously demonstrated by Lorenz.[10] When behaviour implies elements acquired through experience, they are acquired according to a *programme*, and that programme is innate – that is to say, genetically determined. The programme's structure initiates and guides early learning, which will follow a certain pre-established pattern defined in the species' genetic patrimony. This is, no doubt, how we should understand the process whereby the child acquires language. And there is no reason not to suppose that the same holds true for the fundamental categories of cognition in man, and perhaps also for a good many other elements of human behaviour, less basic but of great consequence in the shaping of the individual and society. Such problems are accessible by experiments like those carried out daily by ethologists. Owing to the cruelty of these experiments, their practice upon human beings (in fact, upon young human beings) is unthinkable. Man is thus compelled, through respect for himself, to forego explanation of some of the basic structures of his being.

The lengthy controversy over the Cartesian innateness of

10. K. Lorenz, *Evolution and Modification of Behaviour* (London, Methuen, 1966).

'ideas', denied by the empiricists, is in a way similar to the more recent one which has divided biologists with regard to the distinction between phenotype and genotype. For the geneticists who introduced it the distinction was fundamental, indispensable to the very definition of the hereditary patrimony; but it was very suspect to many biologists not working in genetics, who saw it as a device intended to save the postulate of the invariance of the gene. Here, once more, is a recurrence of the conflict between those for whom truth resides only in the concrete object, actually and fully present, and those who look beyond the object for the ideal form it masks. Alain once said there are only two kinds of scholars: those who love ideas and those who loathe them. In the world of science these two attitudes continue to oppose each other; but both, by their confrontation, are necessary to scientific progress. One can only regret, on behalf of those who scorn ideas, that this progress, to which they contribute, invariably proves them wrong.

In one very important sense, though, the great eighteenth-century empiricists were not wrong. It is perfectly true that in living beings everything, including genetic innateness, comes from experience, whether it be the stereotyped behaviour of bees or the innate framework of human cognition. Everything comes from experience; yet not from actual current experience, reiterated by each individual with each new generation, but instead, from the experience accumulated by the entire ancestry of the species in the course of its evolution. Only this experience wrung from chance – only those countless trials chastened by selection – could, as with any other organ, have made the central nervous system into an organ adapted to its particular function. Only this experience could, in the case of the brain, give a representation of the material world adequate for the performances of the species; furnish a framework permitting efficient classification of the otherwise unusable data of objective experience; and even, in man, simulate experience subjectively so as to anticipate its results and prepare action.

It is the powerful development and intensive use of the simulative function that, in my view, characterize the unique properties of man's brain. And this at the most basic level of the cognitive functions, those on which language rests and which it probably reveals only incompletely.

*the function
of simulation*

Simulation is not an exclusively human function, however. The puppy that shows its joy at seeing its master getting ready for the daily walk obviously imagines – that is, simulates through anticipation – the discoveries it is about to make, the adventures and exciting risks it will face, but without danger thanks to the reassuring presence of its protector. Later on it will simulate the whole thing again, pell-mell, in a dog's dream.

In animals, as in young children too, subjective simulation appears to be only partially dissociated from neuromotor activity. Play is its outward expression. But in man subjective simulation becomes the superior function *par excellence*, the creative function. This is what is reflected by the symbolism of language which, transposing and summarizing its operations, recasts it in the form of speech. Whence the fact, underlined by Chomsky, that even in its humblest employments language is almost always innovative: for it translates a subjective experience, a particular simulation that is always new. And in this too human language is totally unlike animal communication. The latter amounts simply to calls and warnings corresponding to a certain number of stereotyped concrete situations. While doubtless capable of fairly precise subjective simulation, the most intelligent animal has no way 'to unburden its mind' except by roughly indicating in what *direction* its imagination is turned. But man can speak about his subjective experiences: the new experience, the creative encounter, simulated for the first time in a man, can survive and need not be buried with him.

I am sure every scientist must have noticed how his mental reflection, at the deeper level, is not verbal: it is an *imagined experience*, simulated with the aid of forms, of forces, of interactions which together barely compose an 'image' in the visual sense of the term. I have even found myself, after lengthy concentration on the imagined experience to the exclusion of everything else, identifying with a molecule of protein. However, it is not at that moment that the significance of the simulated experience comes clear, but only when it has been enunciated symbolically. Indeed, the nonvisual images with which simulation works should be regarded not as symbols but, if I may so phrase it, as the subjective and abstract 'reality' offered directly to imaginary experience.

However this may be, in everyday practice the process of simulation is entirely masked by the spoken word which follows it almost immediately and which seems inseparable from thought. But, as we know, numerous observations prove that in man the cognitive functions, even the most complex ones, are not immediately linked with speech (nor with any other means of symbolic expression). The studies of various types of aphasia may be cited in particular. Perhaps the most impressive experiments are the recent ones carried out by Sperry on subjects whose two cerebral hemispheres had been separated by surgical cutting of the cross-connecting *corpus callosum*.[11] In these subjects the right eye and right hand communicate information to and receive it from the left hemisphere only. Thus, an object perceived by the left eye or felt with the left hand is recognized without the subject being able to identify it by name. Now in certain difficult tests that involve matching the three-dimensional shape of an object

11. J. Levi-Agresti and R. W. Sperry, *Proceedings of the National Academy of Sciences, 61* (1968), p. 1151.

held in either hand to the flattened-out, two-dimensional picture of that object projected on to a screen, the aphasic right hemisphere proved itself far superior to the 'dominant' left hemisphere – not just more accurate, but able to discriminate more rapidly. It is tempting to speculate upon the possibility that the right hemisphere is responsible for an important part, perhaps the more 'profound' part, of subjective simulation.

If we are correct in considering that thought is based on an underlying process of subjective simulation, we must assume that the high development of this faculty in man is the outcome of an evolution during which natural selection tested the efficacy of the process, its survival value. The very practical terms of this testing have been the success of the concrete action counselled and prepared for by imaginary experimentation. Hence it was on account of its capacity for adequate representation and for accurate foresight *confirmed by concrete experience* that the power of simulation lodged in our early ancestors' central nervous system was propelled to the level reached by *Homo sapiens*. The subjective simulator could not afford to make any mistakes when organizing a panther hunt with the weapons available to Australanthropus, Pithecanthropus or even *Homo sapiens* of Cro-Magnon times. That is why the innate logical instrument we have inherited from our forebears is so reliable and enables us to 'comprehend' events in the world around us, that is, to describe them in symbolic language and to foresee their course, provided the simulator is fed with the necessary elements of information.

As the instrument of intuitive preconception continually enriched by lessons learned from its own subjective experiments, the simulator is the instrument of discovery and of creation. Analysis via language of the logic of its subjective functioning has made possible the formulation of laws of

147

objective logic and the creation of new symbolic instruments such as mathematics. Great thinkers, Einstein among them, have often and rightly marvelled at the fact that the mathematical entities created by man can represent nature so faithfully even though they owe nothing to experience. Nothing, it is true, to individual and concrete experience; but everything to the virtues of the simulator forged by the vast and bitter experience of our humble ancestors. In systematically confronting logic with experience, according to the scientific method, we are in fact confronting all the experience of our ancestors with our own.

We are able to guess the existence of this marvellous instrument, and we know how to translate the result of its operation by language, but we have no idea of its functioning or its structure. Physiological experimentation has so far been unable to help us. Introspection, despite all its dangers, does tell us a little more. There is also the analysis of language, which however only reveals the process of simulation after it has been transformed, and certainly does not reveal all its operations.

There lies the frontier, still almost as impassable for us as it was for Descartes. Until it has been crossed, dualism will continue to be an operative force and truth. Brain and spirit

the dualist illusion and the presence of the spirit

are ideas no more synonymous today than in the eighteenth century. Objective analysis obliges us to see that this seeming duality within us is an illusion; but an illusion so deeply rooted in our being, that it would be vain to hope ever to dissipate it in the immediate awareness of subjectivity, or to learn to live emotionally or morally without it. And, besides, why should one have to? What doubt can there be of the presence of the spirit within us? To give up the illusion that sees in it an immaterial 'substance' is not to deny the existence of the soul, but on the contrary to begin to recognize the complexity, the richness, the unfathomable depth of the genetic and cultural heritage and of the personal experience, conscious or not, which together make up this being of ours, unique and irrefutable witness to itself.

The Kingdom and the Darkness

The day when Australanthropus or one of his kin ventured beyond the communication of actual experience and expressed a subjective experience, a personal 'simulation', saw the birth of a new world, the world of ideas; *the pressures of selection in the evolution of man* and a new evolution, that of culture, became possible. From there on man's physical evolution was to continue for a long time, closely related to the evolution of language, and deeply subjected to its influence, which so changed conditions of selection.

Modern man is the product of that evolutionary symbiosis, and by any other hypothesis incomprehensible, indecipherable. Every living being is *also* a fossil. Within it, all the way down to the microscopic structure of its proteins, it bears the traces if not the stigmata of its ancestry. This is even truer of man than of any other animal species because of the dual evolution – physical and ideational – to which he is heir.

It may be thought that for hundreds of thousands of years ideational evolution barely preceded physical evolution; its progress was hampered by the meagre development of a cortex capable only of anticipating events directly related to immediate survival. Whence the intense selective pressure which was to speed the development of the power of simulation and of the language that conveys its operations.

Whence also the astonishing swiftness of this evolution, to which fossil skulls bear witness.

But as this joint evolution progressed, its ideational component could only tend to greater independence of the restraints which the development of the central nervous system gradually abolished. Owing to this evolution man extended his dominion over the subhuman sphere and suffered less from the dangers it harboured for him. The selective pressure which had guided the first phase of the evolution could then ease, and in any case assumed a different character. Now dominating his environment, man had no serious adversary to face other than his own kind. From then on direct strife between species – war to the death – became one of the principal factors of selection in the human species. In the evolution of animals the phenomenon is extremely rare: today there is no such thing as intraspecific warfare between distinct races or groups in any animal species. Among the larger mammals even single combat, frequent between males, seldom leads to the death of the loser. Specialists all agree in thinking that direct strife, Spencer's 'struggle for life', has played only a minor role in the evolution of species. This is not so as regards mankind. At some point in the development and expansion of the human species tribal or racial warfare came to be an important evolutionary factor. It is quite possible that the sudden disappearance of Neanderthal man was the result of genocide committed by our ancestor *Homo sapiens*. It was not to be the last case of its kind: there are plenty recorded in history.

In what direction did this selective pressure push human evolution? Of course it favoured the expansion of races more generously endowed than others with intelligence, imagination, will and ambition. But it must also have favoured cohesion within the horde, group aggressiveness more than lone courage, respect for the tribal law more than individual initiative.

151

This is a simplified outline, and I am quite willing to accept criticism. I do not mean to divide human evolution into two distinct phases, but have only tried to enumerate the main selective pressures that must have influenced man's cultural and also physical evolution. The important point is that during those hundreds of thousands of years, cultural evolution could not fail to affect physical evolution; in man more than in any other animal – and because of its infinitely greater autonomy – it is *behaviour* that *orients* selective pressure. And once that behaviour ceased to be primarily automatic and became cultural, cultural traits themselves inevitably exerted their pressure upon the evolution of the genome.

This was so until the moment when the accelerating pace of cultural evolution was to mean the genome's complete separation from it.

It is clear that within the framework of modern societies this split is total. In them selection has been done away with. Or at least, there is no longer anything 'natural' about it in the

dangers of genetic degradation in modern societies

Darwinian sense. To the extent that selection is still operative in our societies, it does not favour the 'survival of the fittest' – that is to say, in more modern terms, the genetic

survival of the 'fittest' through a more numerous progeny. Intelligence, ambition, courage, and imagination are certainly still factors of success in modern societies; but of *personal*, not *genetic* success, the only kind that matters for evolution. On the contrary: statistics, as everybody knows, show a negative correlation between the intelligence quotient (or cultural level) and the average number of children per couple. The same statistics also demonstrate that there is a high positive correlation of intelligence quotients between marital partners. This is a dangerous situation, which could gradually drain the highest genetic potential into an élite, shrinking in relative numbers.

This is not all. Until fairly recently, even in relatively 'advanced' societies, the weeding out of both physically and mentally unfit was automatic and ruthless. Most of them did not reach the age of puberty. Today many of these genetic cripples live long enough to reproduce. Thanks to the progress of scientific knowledge and the social ethic, the mechanism which used to protect the species from degeneration (the inevitable result when natural selection is suspended) now hardly functions except in the worst cases.

In confronting these dangers, very often pointed out, we occasionally hear of remedies expected from the current advances in molecular genetics. This illusion, spread about by a few pseudo-scientists, had better be dispelled. It will no doubt be possible to palliate certain genetic flaws, *but only in the afflicted individual*, not in his posterity. Modern molecular genetics offers us *no means whatsoever* for acting upon the ancestral heritage so as to improve it with new features – to create a genetic 'superman'; on the contrary, it reveals the vanity of any such hope: the genome's microscopic proportions today and probably for ever rule out manipulation of this sort. Apart from science fiction's chimerical schemes, the only way of 'improving' the human species would be to introduce deliberate and severe selection. Who would wish, or dare, to do this?

Conditions of nonselection (or of selection-in-reverse) like those reigning in the advanced societies are a definite peril to the species. But to become very serious, however, it would take quite a while: say ten or fifteen generations, or several centuries. And there are far more grave and more urgent dangers already threatening modern societies.

Here I am not referring to the population explosion, to the destruction of the natural environment, nor even to the nuclear stockpile, but to a more insidious and much more deep-seated evil: a sickness of the spirit, the most serious outcome of the ideational evolution which created and ceaselessly worsens it.

The prodigious developments of knowledge over the past three centuries are forcing man to make an agonizing reappraisal of his concept of himself and his relation to the world, a concept rooted in him for tens of thousands of years.

All this, however – the spirit's disorder and the nuclear stockpile alike – comes from one simple idea: that nature is objective, that the systematic confrontation of logic and experience is the sole source of true knowledge. It is hard to understand how, in the kingdom of ideas, this one, so simple and so clear, was not fully grasped until a hundred thousand years after the emergence of *Homo sapiens*; why some of the loftiest civilizations, such as the Chinese, ignored it and had to learn it from the West; or why, in the West itself, it took nearly two thousand five hundred years, from Thales and Pythagoras to Galileo, Bacon, and Descartes, before this idea broke out of its encapsulation within the practice of the mechanical arts.

the selection of ideas

For a biologist it is tempting to compare the evolution of ideas and of the biosphere. For while the abstract kingdom transcends the biosphere by even more than the latter transcends the nonliving universe, ideas have retained some of the properties of organisms. Like these, they tend to perpetuate their structures and to multiply them; they too can fuse, recombine, segregate their content; in short, they too can evolve, and in this evolution selection certainly plays an important role. I shall not hazard a theory of the selection of ideas. But one may at least try to define some of the principal factors involved in it. This selection must necessarily operate at two levels: that of the mind itself and that of performance.

The performance value of an idea depends on the change

it brings to the behaviour of the person or the group that adopts it. The human group upon which a given idea confers greater cohesiveness, greater ambition, and greater self-confidence thereby receives from it an added power to expand which will insure the promotion of the idea itself. Its 'promotion value' bears no relation to the amount of objective truth the idea may contain. The might of the powerful armament provided by a religious ideology for a society does not lie in its structure, but in the fact that this structure is accepted, that it gains command. So one cannot well separate such an idea's power to spread from its power to perform.

The 'spreading power' of an idea is much more difficult to analyse. Let us say that it depends upon pre-existing structures in the mind, among them ideas already implanted by culture, but also undoubtedly upon certain innate structures which are very difficult for us to identify. What is very plain, however, is that the ideas having the highest invading potential are those that *explain* man by assigning him his place in an immanent destiny, a safe harbour where his anxiety dissolves.

For hundreds of thousands of years a man's lot was identical with that of the group, of the tribe he belonged to, and outside which he could not survive. The tribe, for its part, could only survive and defend itself through its cohesion. Whence arose the extreme subjective power of the laws that organized and guaranteed this cohesion. A man might perhaps infringe them; it is unlikely that anyone ever dreamed of denying them. Given the immense selective importance such social structures perforce assumed over such vast stretches of time, it is difficult not to believe that they must have influenced the genetic evolution of the innate categories of the human brain. This evolution

the need for an explanation

155

must not only have facilitated acceptance of the tribal law, but created the *need* for the mythical explanation which gave it a foundation and sovereignty. We are the descendants of these men, and it is probably from them that we have inherited the need for an explanation, the profound disquiet which forces us to search for the meaning of existence. That same disquiet has created all myths, all religions, all philosophies and science itself.

I have very little doubt that this imperious need develops spontaneously, that it is inborn, inscribed somewhere in the genetic code. Apart from the human species, nowhere in the animal kingdom does one find such highly differentiated social organizations except among certain insects: ants, termites, bees. The stability of the social insects' institutions owes next to nothing to cultural heritage, but virtually everything to genetic transmission. Social behaviour, with them, is entirely innate, automatic.

Man's social institutions, which are purely cultural, cannot ever attain such stability; anyway, who would wish for it? The invention of myths and religions, the construction of vast philosophical systems – they are the price man has had to pay in order to survive as a social animal without yielding to pure automatism. But a cultural heritage would not, all alone, have been strong or reliable enough to hold up the social structure. That heritage needed a genetic support to provide something essential to the mind. How else account for the fact that in our species the religious phenomenon is invariably at the base of social structure? How else explain that, throughout the immense variety of our myths, our religions and philosophical ideologies, the same essential 'form' always recurs?

It is easy to see that the 'explanations', which gave a foundation to the law while assuaging man's anxiety, are all 'stories' or, more exactly, 'ontogenies'. Primitive myths almost all tell of more or less divine heroes whose deeds explain the origins of the group and base its social structure upon sacrosanct traditions; one does not remake history. The great religions are of a similar form, based on the story of the life of an inspired prophet who, if not himself the founder of all things, represents that founder, speaks for him, and recounts the history of mankind as well as its destiny. Of all the great religions Judeo-Christianity is probably the most 'primitive' in its strictly historicist structure, being founded on the saga of a Bedouin tribe before being enriched by a divine prophet. Buddhism, which is more highly differentiated, is based in its original form on Karma, the transcending law governing individual destiny. Buddhism is a story of souls rather than of men.

mythic and metaphysical ontogenies

From Plato to Hegel and Marx, the great philosophical systems all propose ontogenies which are both explanatory and normative. It is true that, in Plato's case, the course is downhill rather than ascending. He sees in history only the gradual corruption of ideal forms, and his aim in the *Republic* is to reinstate the past, to move backwards in time.

For Marx, as for Hegel, history unfolds according to an immanent, necessary, and favourable plan. The immense influence of Marxist ideology is not due only to its promise of man's liberation, but also, and probably mainly, to its ontogenic structure, the explanation which it provides, both sweeping and detailed, of past, present, and future history. However, limited to human history, even though decked with the certainties of 'science', historical materialism was still incomplete. It needed the addition of dialectical materialism which provides the total interpretation the mind needs: in this, the history of mankind is bound up with that of the cosmos, obeying the same eternal laws.

If there is an innate need for a complete explanation whose absence causes a deep inner anxiety; if the only form of explanation which can ease the soul is that of a total history which reveals the significance of man by assigning him a necessary place in nature's scheme; if, to appear genuine, meaningful, soothing, the 'explanation' must be fused with the long animist tradition, then we understand why so many thousand years passed before the appearance, in the realm of ideas, of those presenting objective knowledge as the *only* source of real truth.

the breakdown of the old covenant and the modern soul's distress

Cold and austere, proposing no explanation but imposing an ascetic renunciation of all other spiritual fare, this idea could not allay anxiety; it aggravated it instead. It claimed to sweep away at a stroke the tradition of a hundred thousand years, which had become assimilated in human nature itself. It ended the ancient animist covenant between man and nature, leaving nothing in place of that precious bond but an anxious quest in a world of icy solitude. With nothing to recommend it but a certain puritan arrogance, how could such an idea be accepted? It was not; it still is not. If it has commanded recognition, this is solely because of its prodigious powers of performance.

In the course of three centuries, science, founded upon the postulate of objectivity, has won its place in society – in men's practice, but not in their hearts. Modern societies are built upon science. To it they owe their wealth, their power, and the certitude that tomorrow even greater wealth and power will be ours if we so wish. But there is this too: just as an initial 'choice' in the biological evolution of a species can be binding upon its entire future, so the choice of scientific *practice* (an unconscious choice in the beginning) has launched the evolution of culture on a one-way path; on to a track which nineteenth-century scientism saw leading infallibly on to a vast blossoming for mankind, whereas what we see before us today is an abyss of darkness.

Modern societies accepted the treasures and the power offered them by science. But they have not accepted – they have scarcely even heard – its profounder message: the defining of a new and unique source of truth, and the demand for a thorough revision of ethical premises, for a complete break with the animist tradition, the definitive abandonment of the 'old covenant', the necessity of forging a new one. Armed with all the powers, enjoying all the riches they owe to science, our societies are still trying to live by and to teach systems of values already blasted at the root by science itself.

No society before ours was ever torn apart by such conflicts. In both primitive and classical cultures the animist tradition saw knowledge and values stemming from the same source. For the first time in history a civilization is trying to shape itself while clinging desperately to the animist tradition in an effort to justify its values, and at the same time abandoning it as the source of knowledge, of *truth*. The 'liberal' societies of the West still pay lip-service to, and present as a basis for morality, a disgusting farrago of Judeo-Christian religiosity, scientistic progressism, belief in the 'natural' rights of man, and utilitarian pragmatism. The Marxist societies still profess the materialist and dialectical religion of history; on the face of it a more solid moral framework than that of the liberal societies, but perhaps more vulnerable by virtue of the very rigidity which up to now has been its strength. However this may be, all these systems rooted in animism exist outside objective knowledge, outside truth, and are strangers and fundamentally *hostile* to science, which they are willing to use but do not respect or cherish. The divorce is so great, the lie so flagrant, that it can only obsess and lacerate anyone who has some culture or intelligence, or is moved by that moral questioning which is the source of all creativity. It is an affliction, that is to say, for all those who bear or will bear the responsibility for the way in which society and culture will evolve.

The sickness of the modern spirit is this lie at the root of man's moral and social nature. It is this ailment, more or less confusedly diagnosed, that provokes the fear if not the hatred – in any case the estrangement – felt toward scientific culture by so many people today. Their aversion, when openly expressed, is usually directed at the technological by-products of science: the bomb, the destruction of nature, the soaring population. It is easy, of course, to answer that technology and science are not the same thing, and moreover that the use of atomic energy will soon be vital to mankind's survival; that the destruction of nature denotes a faulty technology rather than too much of it; and that the population soars because millions of children are saved from death every year. Are we to go back to letting them die?

This is a superficial reply, confusing the symptoms of the disorder with its underlying cause. For behind the protest is the refusal to accept the essential message of science. The fear is the fear of sacrilege: of outrage to values; and it is wholly justified. It is perfectly true that science attacks values. Not directly, since science is no judge of them and *must* ignore them; but it subverts every one of the mythical or philosophical ontogenies upon which the animist tradition, from the Australian aborigines to the dialectical materialists, has based morality: values, duties, rights, prohibitions.

If he accepts this message in its full significance, man must at last wake out of his millenary dream and discover his total solitude, his fundamental isolation. He must realize that, like a gypsy, he lives on the boundary of an alien world; a world that is deaf to his music, and as indifferent to his hopes as it is to his suffering or his crimes.

Who, then, is to define crime? Who decides what is good and what is evil? All the traditional systems placed ethics and values beyond man's reach. Values did not belong to him;

they were imposed on him, and he belonged to them. Today he knows that they are his and his alone, but now he is master of them they seem to be dissolving in the uncaring emptiness of the universe. It is at this point that modern man turns toward science, or rather against it, now seeing its terrible capacity to destroy not only bodies but the soul itself.

Where is the remedy? Must one claim once and for all that objective truth and the theory of values are eternally opposed, mutually impenetrable domains? This is the attitude adopted *values and knowledge* by many modern thinkers, whether writers, or philosophers, or indeed scientists. I believe that it is not only unacceptable to the vast number of men, whose anxiety it can only perpetuate and worsen; I also believe it is absolutely mistaken, for two essential reasons.

First, of course, because values and knowledge are always and necessarily associated in action as in discourse.

Second, and above all, because *the very definition of 'true' knowledge rests in the final analysis upon an ethical postulate.*

Each of these two points needs to be briefly developed.

Ethics and knowledge are inevitably linked in and through action. Action brings knowledge and values *simultaneously* into play, or into question. All action signifies an ethic, serves or disserves certain values; constitutes a choice of values, or pretends to. On the other hand, knowledge is necessarily implied in all action, while reciprocally, action is one of the two necessary sources of knowledge.

In an animist system the interpenetration of ethics and knowledge creates no conflict, since animism avoids any basic distinction between these two categories: it sees them as two aspects of the same reality. The idea of a social ethic founded upon the so-called 'natural' rights of man also reflects this outlook, displayed, but much more systematically and emphatically, in the attempts to delineate the ethics implicit in Marxism.

From the moment objectivity is made the *conditio sine qua non* of true knowledge, a radical distinction, indispensable to the very search for truth, is established between the domains of ethics and of knowledge. Knowledge in itself is exclusive of all value judgment (except that of 'epistemological value') whereas ethics, in essence *nonobjective*, is for ever barred from the sphere of knowledge.

It is in effect this radical distinction, laid down as an axiom, that created science. I am tempted to suggest that if this unprecedented event in the history of culture occurred in the Christian West rather than in some other civilization, it was perhaps partly thanks to the fundamental distinction drawn by the Church between the domains of the sacred and the profane. Not only did this distinction allow science to pursue its own way (provided it did not trespass on the realm of the sacred); it prepared the mind for the much more radical distinction posed by the principle of objectivity. Westerners often have trouble in understanding that for certain religions there is not and cannot be any distinguishing between sacred and profane: for Hinduism, everything comes within the bounds of the sacred; the very concept of 'profane' is incomprehensible.

But let us return to our main point. The postulate of objectivity, denouncing the 'old covenant', at the same time forbids any confusion of value judgments with judgments arrived at through knowledge. Yet the fact remains that these two categories inevitably unite in the form of action, discourse included. To abide by our principle we shall therefore take the position that no discourse or action is to be considered meaningful, *authentic*, unless – or only insofar as – it makes explicit and preserves the distinction between the two categories it combines. Thus defined, the concept of authenticity becomes the common ground where ethics and knowledge meet again; where values and truth, associated but not interchangeable, reveal their full significance to the attentive man alive to their resonance. In return, *inauthentic* discourse, where the two categories are jumbled, can lead only to the most pernicious nonsense, to the most criminal, even if unconscious, lies.

It is in 'political' discourse (and I mean 'discourse' in the Cartesian sense), of course, that this hazardous amalgamation is most consistently and systematically practised. And not by professional politicians alone. Scientists themselves, outside their field, often prove dangerously incapable of distinguishing between the categories of values and of knowledge.

Animism, we said earlier, neither wants nor for that matter is able to set up an absolute discrimination between value judgments and statements based upon knowledge; for having once assumed that there is an intention, however carefully disguised, present in the universe, what would be the sense of such a distinction? In an objective system the very opposite holds: any confusion of knowledge with values is unlawful, *forbidden*. But – and this is the crucial point – the logical link which radically binds knowledge and values – this ban, this 'first commandment' which ensures the foundation of objective knowledge, itself is not, and cannot be, objective. It is a moral rule, a *discipline*. True knowledge is ignorant of values, but it has to be grounded on a value judgment, or rather on an *axiomatic* value. It is obvious that the positing of the principle of objectivity as the condition of true knowledge *constitutes an ethical choice and not a judgment reached from knowledge, since, according to the postulate's own terms, there cannot have been any 'true' knowledge prior to this arbitral choice.* In order to establish the *norm* for knowledge the objectivity principle defines a *value*: that value is objective knowledge itself. To assent to the principle of objectivity is, thus, to state the basic proposition of an ethical system: *the ethic of knowledge*.

In the ethic of knowledge *it is the ethical choice of a primary value that is the foundation*. The ethic of knowledge thereby differs radically from animist ethics, which all claim to be based on the 'knowledge' of immanent, religious or 'natural' laws which are supposed to impose themselves on man. The ethic of knowledge does not impose itself on man; *on the contrary, it is he who imposes it on himself*, making it the *axiomatic* condition of authenticity for all discourse and all action. The *Discours de la Méthode* proposes a normative epistemology, but it must also be read above all as a moral meditation, a spiritual exercise.

the ethic of knowledge

Authentic discourse is in its turn the foundation of science, and it gives back to man the immense powers that enrich and threaten him today, that free him but might also subjugate him. Modern societies, woven together by science, living from its products, have become as dependent upon it as an addict on his drug. They owe their material power to this fundamental ethic upon which knowledge is based, and their moral weakness to those value-systems devastated by knowledge itself, to which they still try to refer. The contradiction is deadly. This is what is digging the pit we see opening under our feet. The ethic of knowledge that created the modern world is the only ethic compatible with it, the only one capable, once understood and accepted, of guiding its evolution.

Understood and accepted – could it be? If it is true, as I believe, that the fear of solitude and the need for a complete and binding explanation are inborn – that this heritage from the remote past is not only cultural but probably genetic too – can one imagine such an austere, abstract, proud ethic calming that fear, satisfying that need? I do not know. But it may not be altogether impossible. Perhaps, even more than an 'explanation' which the ethic of knowledge cannot supply, man needs to rise above himself, to find transcendence. The abiding power of the great socialist dream, still alive in men's hearts, would indeed seem to suggest it. No system of values can claim to constitute a true ethic unless it proposes an ideal transcending the individual self to the point even of justifying self-sacrifice, if need be.

By the very loftiness of its ambition the ethic of knowledge might perhaps satisfy this craving for something higher. It puts forward a transcendent value, true knowledge, not for the use of man, but for man to serve from deliberate and conscious choice. At the same time it is also a humanist ethic, for it respects man as the creator and repository of that transcendence.

The ethic of knowledge is also in a sense 'knowledge of ethics', that is, of the urges and passions, the needs and limitations of the biological being. It is able to confront the animal in man, to see him not as absurd but strange, precious in his very strangeness: the creature who, belonging simultaneously to the animal kingdom and the kingdom of ideas, is both torn and enriched by this agonizing duality, expressed alike in art and poetry and in human love.

Conversely, the animist systems have to one degree or another preferred to ignore, denigrate or bully biological man, and to make him fear or abhor certain traits inherent in his animal nature. The ethic of knowledge, on the other hand, encourages him to honour and assume this heritage, while knowing how to dominate it when necessary. As for the highest human qualities, courage, altruism, generosity, creative ambition, the ethic of knowledge both recognizes their sociobiological origin and affirms their transcendent value in the service of the ideal it defines.

Finally, the ethic of knowledge is, in my view, the only attitude which is both rational and resolutely idealistic, and on which a real socialism might be built. For the young in spirit that great vision of the nineteenth century still persists with grievous intensity. Grievous because of the betrayals this ideal has suffered, and because of the crimes committed in its name. It is tragic, but was perhaps inevitable, that this profound aspiration had to find its philosophical doctrine in the form of an animist ideology. Looking back, it is easy to see that, from the time of its birth, historical messianism based on dialectical materialism contained the seeds of all the dangers later encountered. Perhaps more than the other animisms, historical materialism is based on a total confusion of the categories of value and knowledge. This very confusion permits it, in a travesty of authentic discourse, to proclaim that it has 'scientifically' established the laws of history, which man has no choice or duty but to obey if he does not wish to sink into oblivion.

the ethic of knowledge and the socialist ideal

This illusion, which is merely puerile when it is not fatal, must be given up once and for all. How can an authentic socialism ever be built on an essentially inauthentic ideology, a caricature of that very science whose support it claims (most sincerely, in the minds of its followers)? Socialism's one hope is not in a 'revision' of the ideology that has been dominating it for over a century, but in completely abandoning that ideology.

Where then shall we find the source of truth and the moral inspiration for a really *scientific* socialist humanism? Only, we suggest, in the sources of science itself, in the ethic upon which knowledge is founded, and which by free choice makes knowledge the supreme value – the measure and guarantee for all other values. An ethic which bases moral responsibility upon the very freedom of that axiomatic choice. Accepted as the foundation for social and political institutions, and as the measure of their authenticity and their value, only the ethic of knowledge could lead to socialism. It prescribes institutions dedicated to the defence, the extension, the enrichment of the transcendent kingdom of ideas, of knowledge, and of creation – a kingdom which is within man, where progressively freed both from material constraints and from the misleading servitudes of animism, man could at last live authentically; there he would be protected by institutions which, seeing him as both the subject of the kingdom and its creator, would serve him in his unique and precious essence.

This is perhaps a utopia. But it is not an incoherent dream. It is an idea that owes its strength to its logical coherence alone. It is the conclusion to which the search for authenticity necessarily leads. The ancient covenant is in pieces; man at last knows that he is alone in the unfeeling immensity of the universe, out of which he emerged only by chance. Neither his destiny nor his duty have been written down. The kingdom above or the darkness below: it is for him to choose.

Appendices

1 Structure of Proteins

Proteins are macromolecules constituted by the linear polymerization of compounds called amino acids. The general structure of the 'polypeptide' chain resulting from this polymerization is shown in the following drawing.

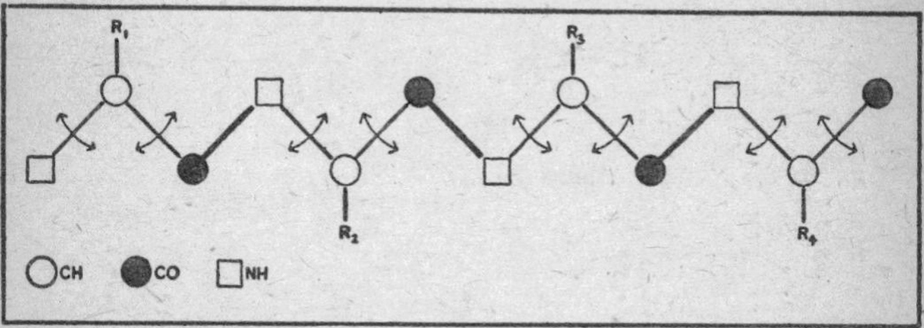

The white and black circles and the white squares correspond to various groupings of atoms (O$=$CH; ●$=$CO; □$=$NH), while the letters R_1, R_2, etc. represent different organic residues. The twenty amino acid residues that are the universal constituents of proteins are shown in Table 1.

The chain includes three kinds of bonds between atoms or groups of atoms.

1. Between white circle and black circle (CH—CO);
2. Between white circle and white square (CH—NH);
3. Between black circle and white square (CO—NH).

The last bond, known as the 'peptide bond' (represented by heavy lines in the drawing above), is rigid: the atoms it connects are held immobile with respect to each other. The two other bonds, on the contrary, permit the linked atoms to

Table 1 *Amino Acid Residues*

A. HYDROPHOBIC

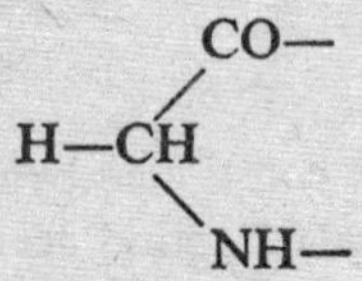

$$CO-$$
$$H-CH$$
$$NH-$$

Glycyl (GLY)

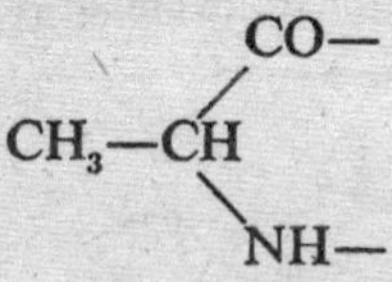

$$CO-$$
$$CH_3-CH$$
$$NH-$$

Alanyl (ALA)

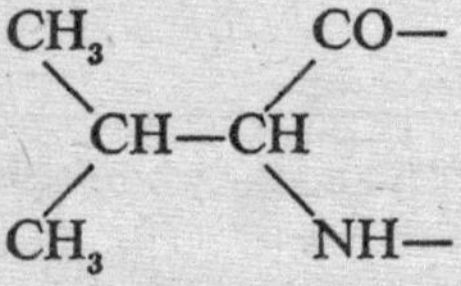

$$CH_3$$
$$CH-CH$$
$$CH_3 \qquad NH-$$
$$CO-$$

Valyl (VAL)

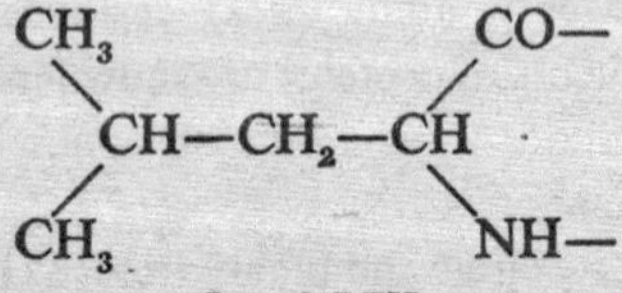

$$CH_3$$
$$CH-CH_2-CH$$
$$CH_3 \qquad NH-$$
$$CO-$$

Leucyl (LEU)

$$CH_3-CH_2 \qquad CO-$$
$$CH-CH$$
$$CH_3 \qquad NH-$$

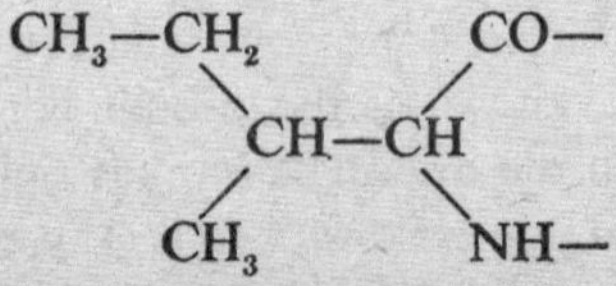

Isoleucyl (ILEU)

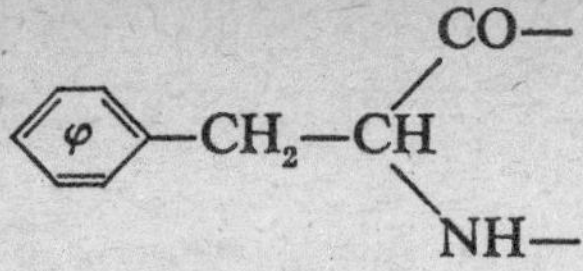

Phenylalanyl (PHE)

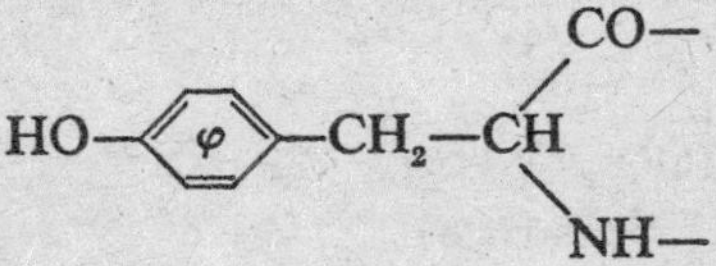

Tyrosyl (TYR)

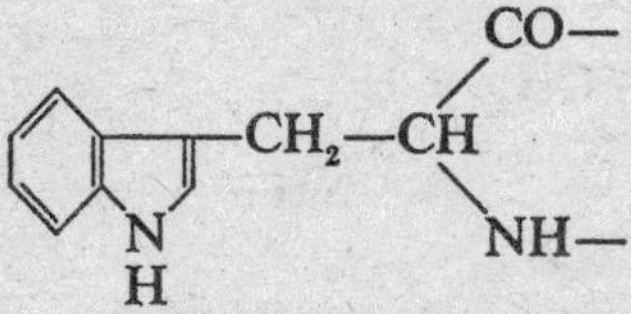

Tryptophanyl (TRY)

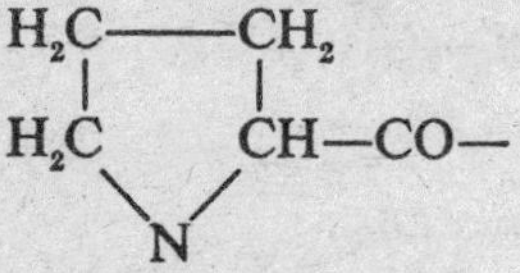

Prolyl (PRO)

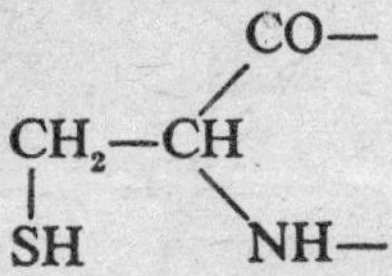

Cysteinyl (CYS)

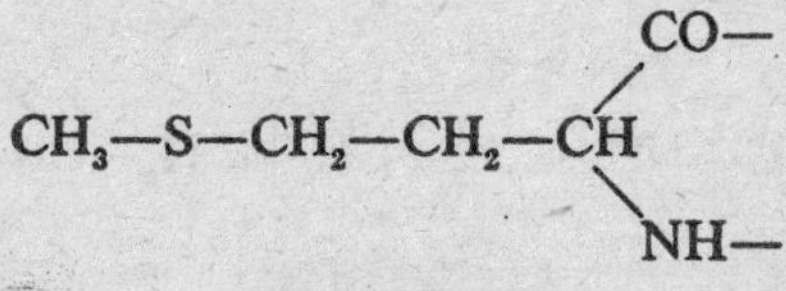

Methionyl (MET)

Table 1 *Amino Acid Residues*

B. HYDROPHILIC

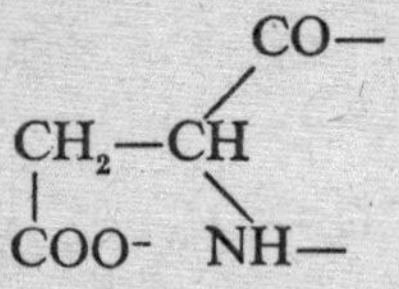

$$CH_2-CH \begin{array}{c} CO- \\ \\ NH- \end{array}$$

$$COO^-$$

Aspartyl (ASP)

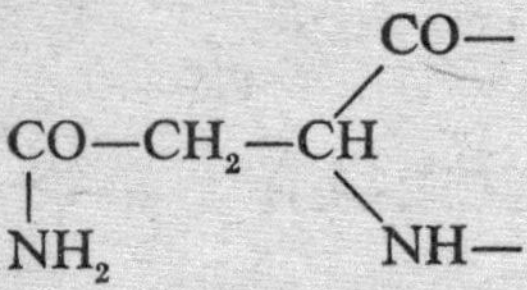

$$CO-CH_2-CH \begin{array}{c} CO- \\ \\ NH- \end{array}$$

$$NH_2$$

Asparagyl (ASPN)

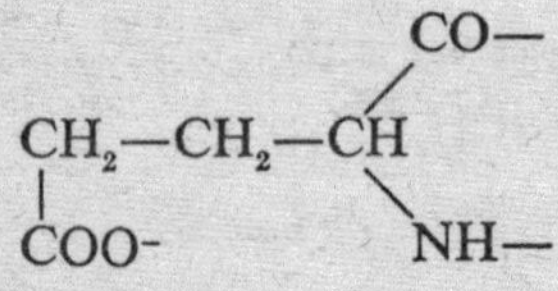

$$CH_2-CH_2-CH \begin{array}{c} CO- \\ \\ NH- \end{array}$$

$$COO^-$$

Glutamyl (GLU)

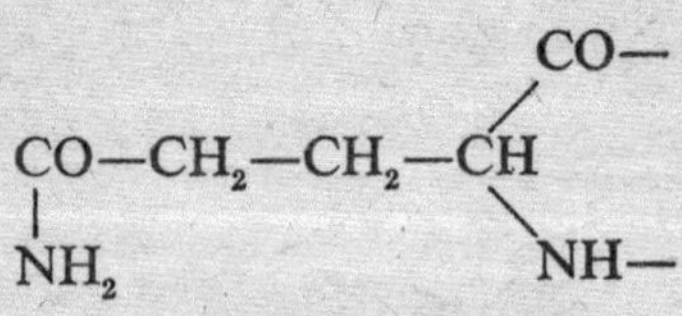

$$CO-CH_2-CH_2-CH \begin{array}{c} CO- \\ \\ NH- \end{array}$$

$$NH_2$$

Glutaminyl (GLUN)

$$\begin{array}{c} HN \\ \\ C-NH-CH_2-CH_2-CH_2-CH \\ \\ H_3^+N \end{array} \begin{array}{c} CO- \\ \\ NH- \end{array}$$

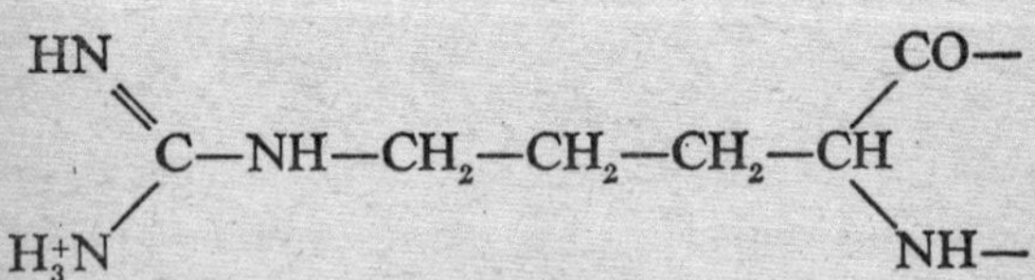

Arginyl (ARG)

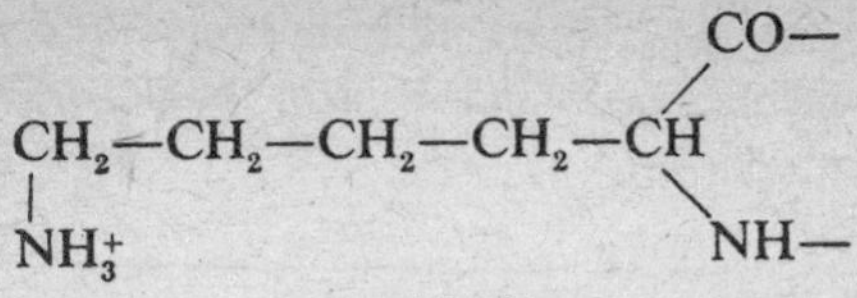

$$CH_2-CH_2-CH_2-CH_2-CH\overset{\displaystyle CO-}{\underset{\displaystyle NH-}{\big|}}$$

$$\overset{\displaystyle |}{NH_3^+}$$

Lysyl (LYS)

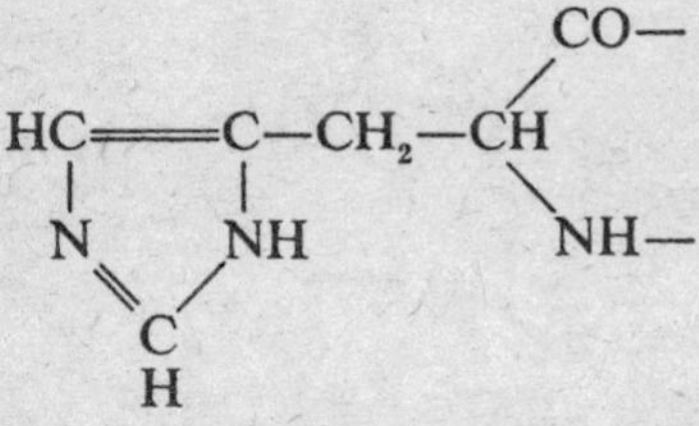

$$HC=C-CH_2-CH\overset{\displaystyle CO-}{\underset{\displaystyle NH-}{\big|}}$$

$$\overset{\displaystyle |}{N}\quad \overset{\displaystyle |}{NH}$$

$$\overset{\displaystyle C}{\underset{\displaystyle H}{\big|}}$$

Histidyl (HIS)

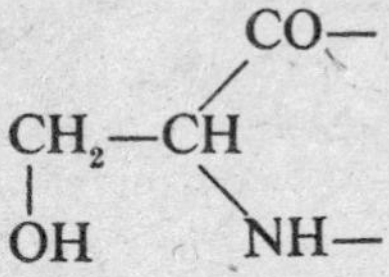

$$CH_2-CH\overset{\displaystyle CO-}{\underset{\displaystyle NH-}{\big|}}$$

$$\overset{\displaystyle |}{OH}$$

Seryl (SER)

$$CH_3-CH-CH\overset{\displaystyle CO-}{\underset{\displaystyle NH-}{\big|}}$$

$$\overset{\displaystyle |}{OH}$$

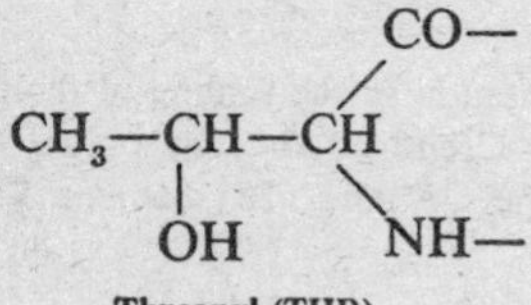

Threonyl (THR)

rotate with respect to each other (curved arrows in the drawing). This in turn permits the polypeptide fibre to bundle by folding in an extremely varied and flexible way. Only the room taken up by the atoms (notably those that constitute the residues R_1, R_2, etc.) places any limitation upon the fibre's possibilities for folding.

However (see p. 90), in native globular proteins all the molecules of a given chemical species (defined by the *sequence* of the residues in the chain) adopt the same bundled shape.

Figure 5, below, diagrammatically illustrates the complex and seemingly incoherent arrangement of the polypeptide chain in the enzyme papain.

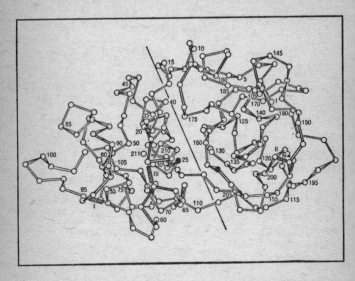

Figure 5. Schematic representation of the folding of the peptide chain in the papain molecule. J. Drenth, J. N. Jansonius, R. Koekoek, H. M. Swen, and B. G. Wolthers, *Nature, 218* (1968), pp. 929–32.

2 Nucleic Acids

Nucleic acids are macromolecules resulting from the linear polymerization of compounds called 'nucleotides'. The latter are formed through the association of a sugar with a nitrogen-containing base on the one hand, and on the other with a phosphoryl group. The polymerization occurs through the intermediary of phosphoric groups which link each sugar residue to the one before and the one after, thus forming a 'polynucleotide' chain.

In DNA (deoxyribonucleic acid) are found four nucleotides which differ in the structure of the constituent nitrogen-containing base. These four bases, adenine, guanine, cytosine, and thymine, are usually abbreviated as A, G, C, and T. They are the letters of the genetic alphabet. For steric reasons the adenine (A) in DNA tends to form a spontaneous noncovalent association (see p. 59) with thymine (T), while guanine (G) associates with cytosine (C).

DNA is made up of *two* polynucleotide strands joined by means of these specific noncovalent bonds. In the double strand, A of one strand is linked to T in the other; G to C; T to A; and C to G. The two strands are therefore *complementary*.

This structure is represented diagrammatically in the figure following. In it the pentagons stand for sugar residues, the black circles for the phosphorus atoms which ensure the continuity of both chains; the squares marked A, T, G, and C represent the bases matched in pairs (A–T; G–C;

T–A; C–G) by noncovalent interactions, indicated by the dotted lines. The structure can accommodate every possible sequence of pairs. It is not limited as to length.

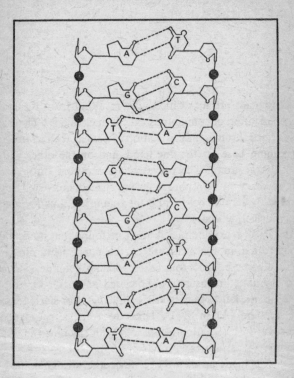

The *replication* of this molecule proceeds by the separation of the duplex, followed by the reconstitution, nucleotide by nucleotide, of the two complements. This is shown – in a simplified manner and confining ourselves to four pairs – in the drawing following.

The two molecules thus synthesized each contain one strand of the parent molecule and a strand newly formed by specific nucleotide-by-nucleotide pairing. These two molecules

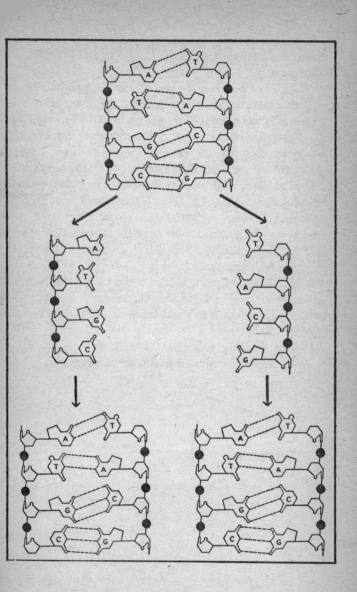

179

are identical to each other and to the original molecule. Such is the mechanism, very simple in principle, of replicative invariance.

Mutations result from the various kinds of accidents which may alter this microscopic mechanism. The chemistry behind some mutations is today fairly well understood. For example, the substitution of one nucleotide pair for another is accounted for by the fact that the nitrogen-containing bases, besides their 'normal' state, can exceptionally and temporarily adopt a form in which their capacity for specific base-pairing is so to speak 'reversed' (thus, in its 'exceptional' state, base C pairs with A rather than with G, and so on). Chemical agents are known which considerably augment the probability, that is to say the frequency, of these 'illicit' pairings. These agents are powerful 'mutagens'.

Other chemical agents, able to wedge themselves *between* the nucleotides in a DNA strand, deform it and thereby induce such accidents as the deletion or addition of one or several extra nucleotides.

Finally, ionizing radiations (X-rays and cosmic rays) provoke, *inter alia*, various deletions or 'garblings'.

3 The Genetic Code

The structure and properties of a protein are defined by the sequence (the linear order) of the amino acid residues in the polypeptide (cf. p. 91). This sequence is itself determined by that of the nucleotides in a segment of DNA strand. The genetic code (*sensu stricto*) is the rule which prescribes, given polynucleotide sequence, the corresponding polypeptide sequence.

Since there are twenty amino acids to specify and at the same time only four 'letters' (four nucleotides) in the DNA alphabet, several nucleotides are required for the specifying of each amino acid. The code in fact reads in 'triplets': each amino acid is specified by a sequence of *three nucleotides*. The general features of the code are summarized in Table 2, on p. 182.

It is to be noted at once that the translation machinery does not make direct use of the DNA nucleotide sequences themselves but of a working copy formed by the 'transcription' of *one* of the two strands into a one-stranded polynucleotide called 'messenger ribonucleic acid' (messenger RNA). The RNA polynucleotides differ from the DNA nucleotides in a few details of structure, notably the substitution of the base uracil (U) for the base thymine (T). Since messenger RNA serves directly as template for the sequential assembly of the amino acids which are to make up the polypeptide, the code, shown in Table 2, is here written out in the RNA rather than the DNA alphabet.

Table 2 *The Genetic Code*

I	II	U	C	A	G	III
		PHE	SER	TYR	CYS	U
	U	PHE	SER	TYR	CYS	C
		LEU	SER	nonsense	nonsense	A
		LEU	SER	nonsense	TRY	G
		LEU	PRO	HIS	ARG	U
	C	LEU	PRO	HIS	ARG	C
		LEU	PRO	GLUN	ARG	A
		LEU	PRO	GLUN	ARG	G
		ILEU	THR	ASPN	SER	U
	A	ILEU	THR	ASPN	SER	C
		ILEU	THR	LYS	ARG	A
		MET	THR	LYS	ARG	G
		VAL	ALA	ASP	GLY	U
	G	VAL	ALA	ASP	GLY	C
		VAL	ALA	GLU	GLY	A
		VAL	ALA	GLU	GLY	G

In this table the first letter for each triplet is read in the vertical column on the left; the second letter in the horizontal row at the top; the third in the vertical column on the right. The names of the amino acid residues are given in the abbreviations indicated in Table 1, pages 172–5.

We see that for most of the amino acids there exist several different notations in the form of nucleotide 'triplets'. With a four-letter alphabet $4^3 = 64$ three-letter 'words' can be formed; there are however only 20 residues to be specified.

On the other hand three triplets (UAA, UAG, UGA) are labelled 'nonsense' because they do not designate any amino acid. They do nevertheless play an important role as punctuation signals (at the beginning or end) in reading the nucleotide sequence.

The actual mechanism of translation is complex; numerous macromolecular constituents are involved in it. A familiarity with this mechanism is not indispensable to an understanding of the text. It will be enough to say a few words about the intermediates that hold the key to the translation process. These intermediates are the so-called 'transfer' RNA molecules. These contain:

1. A group which 'accepts' amino acids; special enzymes

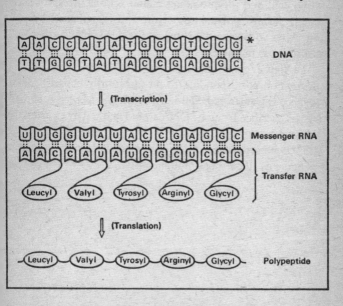

recognize, on the one hand an amino acid, on the other hand a particular transfer RNA, and catalyse the covalent association of the amino acid with the RNA molecule.

2. A sequence complementary to each of the code's triplets, which enables each transfer RNA to pair with the corresponding triplet of messenger RNA.

The pairing comes about in association with a complex constituent, the ribosome, as it were the 'workbench' upon which the various components of the mechanism are put together. The messenger RNA is read sequentially, an as yet imperfectly understood mechanism permitting the ribosome to move, triplet by triplet, along the polynucleotide chain. In its turn each triplet pairs on the surface of the ribosome with the corresponding messenger RNA carrying the amino acid specified by that triplet. At each stage an enzyme catalyses the formation of a peptide bond between the RNA-borne amino acid and the preceding amino acid at the end of the already formed polypeptide chain, thus lengthened by one unit. After which the ribosome moves one triplet further and the process is repeated.

The figure above outlines the mechanism whereby the information corresponding to an (arbitrarily chosen) DNA sequence is transferred. Here the messenger RNA is assumed to be transcribed from the DNA strand marked by an asterisk. In actual practice the transfer RNAs pair one after another with the messenger; for the sake of clarity, they are shown in this figure as all pairing simultaneously.

4 Note Concerning the Second Law of Thermodynamics

So much has been written on the meaning of the second law, on entropy, on the 'equivalence' between negative entropy and information, that one hesitates to review this subject in a few brief paragraphs. One or two general remarks may however prove of use to some readers.

In the form originally put forward (by Clausius in 1850, as a generalization of Carnot's principle), the second law specifies that *within an energetically isolated enclosure* all differences of temperature must tend to even out *spontaneously*. Or again – and it comes to the same thing – within such a space, if the temperature is *uniform* to begin with, no differences of thermal potential can possibly appear in different areas of the whole. Whence the necessity to expend energy in order to cool a refrigerator, for example.

Now within an insulated and enclosed space at uniform temperature, where no difference of potential remains, no (macroscopic) phenomenon can occur. The system is *inert*. In this sense we say that the second law specifies the inevitable *degradation* of energy within an isolated system, such as the universe. 'Entropy' is the thermodynamic quantity which measures the extent to which a system's energy is thus degraded. Consequently, according to the second law every phenomenon, whatever it may be, is necessarily accompanied by an increase of entropy within the system where it occurs.

It was the development of the kinetic theory of matter (or statistical mechanics) that brought out the deeper and broader significance of the second law. The 'degradation of energy' or the increase of entropy is a statistically predictable consequence of the random movements and collisions of molecules. Take for example two enclosed spaces at different temperatures put into communication with each other. The 'hot' (i.e., fast) molecules and the 'cold' (slow) molecules will, in the course of their movements, pass from one space into the other, thus eventually and inevitably nullifying the temperature difference between the two enclosures. From this example one sees that the increase of entropy in such a system is linked to an increase of *disorder*: the fast and the slow molecules, at first separate, are now intermingled, and the total energy of the system will distribute statistically among them all as a result of their collisions; what is more, the two enclosures, at first discernibly different (in temperature) now become equivalent. Before the mixing, work could be accomplished by the system, since it involved a difference of potential between the enclosures. Once statistical equilibrium is achieved within the system, no further macroscopic phenomenon can occur there.

If increased entropy in a system spells out a commensurate increase of *disorder* within it, an increase of order corresponds to a diminution of entropy or, as it is sometimes phrased, a heightening of negative entropy (or 'negentropy'). However, the degree of order in a system is definable under certain conditions in another language: that of information. The order of a system, in such terms, is equal to the quantity of information required for the *description* of that system. Whence the idea, propounded by Szilard and Léon Brillouin, of a certain equivalence between 'information' and 'negentropy' (see p. 64). An exceedingly fertile idea; but one which may give rise to ambiguous generalizations or assimilations.

Nevertheless it is legitimate to regard one of the fundamental statements of information theory, namely that the transmission of a message is necessarily accompanied by a certain dissipation of the information it contains, as the theoretical equivalent of the second law of thermodynamics.

Acknowledgments

The translator and publishers wish to acknowledge their indebtedness for permission to reproduce copyright material from the following books: *The Myth of Sisyphus* by Albert Camus, translated by Justin O'Brien, Copyright © 1955 The Estate of Albert Camus, Hamish Hamilton, London; *Dialectics of Nature* by Frederick Engels, translated by Clemens Dutt, and published by Lawrence and Wishart, London, 1940; *Herr Eugen Dühring's Revolution in Science (Anti-Dühring)* by Frederick Engels, translated by Emile Burns, and published by Martin Lawrence, London, 1935.

Also available in Fount Paperbacks

Memories, Dreams, Reflections
C. G. JUNG

'Jung's single-minded humility, his passion to unearth truth, is one of the loveliest impressions to emerge from this absorbing and many-sided book.'   *The Times*

Myths, Dreams and Mysteries
MIRCEA ELIADE

'A penetrating and sympathetic scrutiny of the mythologies that vivify ancient communities and tell us so much of the perennial meaning and destiny of man.'   *The Times Literary Supplement*

Silent Music
WILLIAM JOHNSTON

Silent Music is a brilliant synthesis which joins traditional religious insights with the discoveries of modern science to provide a complete picture of mysticism – its techniques and stages, its mental and physical aspects, its dangers, and its consequences.

The Varieties of Religious Experience
WILLIAM JAMES

'A classic of psychological study . . . fresh and stimulating . . . this book is a book to prize.'   *The Psychologist*

The Religious Experience of Mankind
NINIAN SMART

'Professor Smart's patient, clear and dispassionate exposition makes him a tireless and faithful guide.'   *Evening News*

Fount Paperbacks

Fount is one of the leading paperback publishers of religious books and below are some of its recent titles.

- ☐ DISCRETION AND VALOUR (New edition)
  Trevor Beeson £2.95 (LF)
- ☐ ALL THEIR SPLENDOUR David Brown £1.95
- ☐ AN APPROACH TO CHRISTIANITY
  Bishop Butler £2.95 (LF)
- ☐ THE HIDDEN WORLD Leonard Cheshire £1.75
- ☐ MOLCHANIE Catherine Doherty £1.00
- ☐ CHRISTIAN ENGLAND (Vol. 1)
  David Edwards £2.95 (LF)
- ☐ MERTON: A BIOGRAPHY Monica Furlong £2.50 (LF)
- ☐ THE DAY COMES Clifford Hill £2.50
- ☐ THE LITTLE BOOK OF SYLVANUS
  David Kossoff £1.50
- ☐ GERALD PRIESTLAND AT LARGE
  Gerald Priestland £1.75
- ☐ BE STILL AND KNOW Michael Ramsey £1.25
- ☐ JESUS Edward Schillebeeckx £4.95 (LF)
- ☐ THE LOVE OF CHRIST Mother Teresa £1.25
- ☐ PART OF A JOURNEY Philip Toynbee £2.95 (LF)

All Fount paperbacks are available at your bookshop or newsagent, or they can also be ordered by post from Fount Paperbacks, Cash Sales Department, G.P.O. Box 29, Douglas, Isle of Man, British Isles. Please send purchase price, plus 10p per book. Customers outside the U.K. send purchase price, plus 12p per book. Cheque, postal or money order. No currency.

NAME (Block letters) _____

ADDRESS _____

_____

_____